Food Engineering Series

Series Editor

Gustavo V. Barbosa-Cánovas, Washington State University, Pullman, WA, USA

Advisory Editors

José Miguel Aguilera, Catholic University, Santiago, Chile

Kezban Candoğan, Ankara University, Ankara, Turkey

Richard W. Hartel, University of Wisconsin, Madison, USA

Albert Ibarz, University of Lleida, Lleida, Spain

Micha Peleg, University of Massachusetts, Amherst, MA, USA

Shafiur Rahman, Sultan Qaboos University, Al-Khod, Oman

M. Anandha Rao, Cornell University, Geneva, NY, USA

Yrjö Roos, University College Cork, Cork, Ireland

Jorge Welti-Chanes, Tecnológico de Monterrey, Monterrey, Mexico

Springer's *Food Engineering Series* is essential to the Food Engineering profession, providing exceptional texts in areas that are necessary for the understanding and development of this constantly evolving discipline. The titles are primarily reference-oriented, targeted to a wide audience including food, mechanical, chemical, and electrical engineers, as well as food scientists and technologists working in the food industry, academia, regulatory industry, or in the design of food manufacturing plants or specialized equipment.

More information about this series at https://link.springer.com/bookseries/5996

Keshavan Niranjan

Engineering Principles for Food Process and Product Realization

 Springer

Keshavan Niranjan
Department of Food and Nutritional Sciences
University of Reading
Reading, Berkshire, UK

ISSN 1571-0297
Food Engineering Series
ISBN 978-3-031-07572-8   ISBN 978-3-031-07570-4   (eBook)
https://doi.org/10.1007/978-3-031-07570-4

This Springer imprint is published by the registered company Springer Nature Switzerland AG
The registered company address is: Gewerbestrasse 11, 6330 Cham, Switzerland

To
Sai,
Atmika, Aaditya
Robert, Arun and Ayush

Preface

This book aims to serve as an introduction to the core underpinning principles of food engineering – a subject which has struggled to find a clear identity of its own over the years, instead operating as a subset of other branches of engineering, principally chemical and agricultural engineering. There is no doubt that food engineering is allied with chemical engineering because food systems can be considered akin to biochemical systems that are somewhat weakly characterised. But there are stark differences in formulation, processing, storage and consumption of foods – which, whilst distinguishing it from other branches of engineering, also give food engineering a unique identity as an academic and practising discipline. The number of processing operations practised in food industry, to produce a plethora of food products for a culturally diverse world, far exceeds the number of operations practised in the conventional chemical sector (Niranjan 1994). More importantly, the products are consumed out of choice and free will, with hardly any constraining regulations on the quantities consumed, which makes it necessary for food engineers to understand not only the factors governing choice, liking, and sensory impact but also the fate of food in the gastro-intestinal tract (GIT) along with its short and longer-term health impact. This requirement places on the subject a strong responsibility to contribute to the society's health and well-being, in addition to environment and sustainability. With health, sustainability and security being the main drivers of food engineering as an academic discipline, a reconceptualised definition of the subject was recently proposed: *Food Engineering is the work of designing, formulating and manipulating food products which have desired sensory, satiety, health and well-being responses, and developing – across various operational scales – designs for the lowest environmental impact processing, packaging and storage systems which are capable of realising the products and attributes* (Niranjan 2016). Consistent with this definition, there is also a concomitant need to develop resources that would train students and practitioners of food engineering to become competent in quantitative process and product design. As of now, this can only be done by referring to books on other engineering disciplines and adapting these to the unique requirements of food systems.

This book is therefore dedicated to food systems and identifies the key quantitative principles that one must become familiar with, before embarking upon learning detailed process and product design – which this book will hopefully appetize. The ten chapters of this book therefore expand on relevant principles but selectively use common food processes to illustrate their applications. The chapters cover topics which are often referred to as process engineering science, such as mass and energy balances, flow of fluids, heat transfer, mass transfer, thermal processing, reaction kinetics, and thermodynamics. In addition, the book also covers: (i) the engineering principles of the movement and transformation of food through the gastro-intestinal tract, (ii) the basic principles of water use and wastewater handling in food processing, (iii) the assessment of environmental impact of food product manufacture, and (iv) key engineering principles applied to product development such as conceiving flow sheet, establishing economic viability and the basics of project management – all of which will be necessary to realise a product in quality, quantity and efficacy – which is the ultimate goal of studying food engineering as a major subject.

The text within each chapter assumes that the reader is familiar with basic mathematics, physics and chemistry, normally covered at a pre-university level. Each chapter has also been developed and presented in such a way that it is concise, self-contained, and not excessively reliant on literature published earlier. As such, the list of references at the end of each chapter is short. However, references for further reading have been included at the end of each chapter to continue building on the knowledge gained from this book. Each topic has also been illustrated with solved problems, so that the application and understanding of the principles studied get reinforced. Further, a set of problems are included at the end of the book which serve to test a reader's understanding of each chapter; the answers to these problems are also included as supplementary material.

I must acknowledge that the content and presentation of this book has been profoundly influenced by my teaching experience of over three decades at the University of Reading (UK) and has inevitably been informed by the interactions I have had with colleagues at Reading and all over the world – too numerous to list all the individuals here separately, but each richly deserving of my heartfelt gratitude. I would however mention my mentor, the Late Professor Leo Pyle, who despite having passed away over a decade ago continues to inspire me, and indeed did so to write this book. I would also like to acknowledge with immense gratitude the help rendered by Professor Rekha Singhal (ICT Mumbai), Dr Nikos Mavroudis (University of Reading), Ms Aratrika Ray (ICT Mumbai) and Mr Rahul Kumar (University of Reading) in giving this book the finishing touches it needed.

It is hoped that students and teachers of food engineering and related areas will embrace this book as an introductory text within their programmes.

Reading, UK   Keshavan Niranjan
February 2022

References

Niranjan K (1994) Chemical engineering principles and food processing. Trends Food Sci Technol 5(1):20–23

Niranjan K (2016) A possible reconceptualization of food engineering discipline. Food Bioprod Process 99:78–89

Contents

Chapter 1
Mass and Energy Balances

Keshavan Niranjan

Aim Material and Energy requirements form the heart of any process, and this chapter shows how these requirements can be calculated around single operations and complete processes.

1.1 Introduction

Raw materials are converted to finished products within a *process*. Most, if not all, manufacturing operations are carried out in a series of processing steps, known more commonly in the chemical industry as *unit operations*. In the context of food manufacture, this term is somewhat inaccurate because each *operation* can bring about a multitude of effects on food, some desirable, and others less so. Typical operations encountered in food manufacture include mixing, separations, reactions, heat transfer, packaging etc. It is impossible to list all the operations practised by food industries, but, in terms of numbers, it is an order of magnitude greater than the *unit operations* practised by the chemical industry. One possible classification of food processing operations is given in Table 1.1.

K. Niranjan
Department of Food and Nutritional Sciences, University of Reading, Reading, Berkshire, UK
e-mail: afsniran@reading.ac.uk

© Springer Nature Switzerland AG 2022
K. Niranjan, *Engineering Principles for Food Process and Product Realization*,
Food Engineering Series, https://doi.org/10.1007/978-3-031-07570-4_1

Table 1.1 Classification of Food Processing Operations. (Modified from Niranjan 1994)

1	**Preliminary Processes**	
	Classification	Sorting, Grading
	Surface processes	Cleaning, degumming, peeling, deskinning, defeathering, dehusking, podding, shelling, silking, hulling, coating, scarifying, rendering
	Others	Butchering, evisceration, filleting, coring, pitting, trimming, stemming
2a	**Conversion-physical**	
	Size reduction	Cutting, slicing, dicing, grating/shredding, milling, pulverising (grinding), emulsification, homogenization
	Size enlargement	Agglomeration, granulation, instantization, coagulation, flocculation, pelleting
	Reshaping	Sheeting, rounding, flaking, puffing
	Mixing operations	Blending, dispersing, aerating/deaerating (carbonating, whipping, beating), kneading
	Mass transfer operations	Extraction, leaching, distillation, rehydration
	Others	Expressing, encapsulating
2b	**Conversion-Chemical**	
	Living systems	Fermentation
	Non-living biological systems	Enzymatic processes
	Chemical reactions	Acidification (pickling), hydrolysis, sulphating, caramelization
2c	**Conversion-physicochemical**	
	Cooking, baking, roasting, frying, extrusion, smoking	
3	**Preservation techniques**	
	Elevated temperature processes[a]	Blanching, pasteurisation, sterilisation, drying/dehydration, concentration
	Ambient temperature and the so-called non-thermal processes	Salting/brining, fumigation, irradiation, high pressure processing, pulsed electric field processing, membrane concentration
	Low temperature processes[a]	Chilling, freezing, freeze concentration, freeze drying
	Packaging	Canning, bottling, flexible pouches, coating (waxing) made from metals, glass, paper, plastic and other materials
4	**Separation techniques**	
	Crystallisation, precipitation, clarification, filtration, centrifugation, evaporation, membrane processes,	

[a]Methods for adding or taking away heat include indirect heat transfer (where the food does not come into contact with the heating or cooling medium) and direct heat transfer (where the food makes direct contact with the heating/cooling medium. In addition, the food can also be made to generate heat from other forms of energy such as microwaves or electricity (ohmic heating)

1.2 Definition of Food Composition

A food manufacturing process generally involves a number of streams which flow in and out of equipment. These streams are invariably *multi-component*, and often *multi-phase*. It is therefore critical to specify the composition of such streams. There are a number of ways in which this can be done, but the two most common ways are to specify *Mass fraction* (or percentage) and *Molar fractions* (or percentage).

If m_A m_B and m_C are the masses of components A, B and C, respectively, the mass fraction of A is given by:

$$x_A = \frac{m_A}{m_A + m_B + m_C} \quad ; \tag{1.1}$$

and its percentage is: $100x_A$. Likewise, the fractions and percentages of other components can also be written down. It is necessary to note that the sum of all fractions must add to unity; and that of all percentages to 100.

When the chemical composition of each component is clearly known (and this is often not the case in food manufacturing practice), it is more convenient to work with *Molar units*, especially when chemical reactions are taking place in the process. The normal molar unit used in processing practice is kmol and it is defined as:

$$number\ of\ kmols = \frac{mass\ of\ the\ component\ in\ kg}{molecular\ mass}$$

Compositions can be expressed in terms of *mole fractions* or *mole percentages* just as in Eq. (1.1). In other words, the mole fraction of A in the mixture is:

$$X_A = \frac{m_A/M_A}{m_A/M_A + m_B/M_B + m_C/M_C} \tag{1.2}$$

where M_A, M_B and M_C are the molecular masses of A, B and C, respectively.

1.2.1 Moisture Content

Water is the key component of, virtually, all biological materials and *moisture content* can be expressed in two ways: one as a mass fraction (also known as *wet basis*) and the other as a mass ratio (known as *dry basis*); in either case, the amounts of all other components (i.e. non-aqueous) in the material are lumped into one and refereed to as the *dry matter* content. Thus, if m_w and m_S are the masses of water and dry matter respectively, the moisture content on a wet-basis is:

$$X_{wb} = \frac{m_W}{m_W + m_S} \tag{1.3}$$

The moisture content on a dry basis, on the other hand, is given by:

$$X_{db} = \frac{m_W}{m_S} \tag{1.4}$$

It follows from Eqs. (1.3) and (1.4) that X_{wb} ranges from 0–1 (or 0–100 if expressed as percent), whereas X_{db} ranges from zero to infinity. Moreover, if the moisture content on a dry basis is known, that on a wet basis can be calculated, and *vice versa*, using:

$$X_{wb} = \frac{X_{db}}{1 + X_{db}} \tag{1.5}$$

1.2.2 Composition of Gaseous Mixtures

When gaseous mixtures are being considered, their composition is often expressed in terms of *partial pressures*, which is defined as the pressure exerted by a given component if it had occupied the entire volume of the mixture at the same temperature. It follows that the sum of the partial pressure of each component must add up to the total pressure; this is commonly known as the *Dalton's law*. When the gas phase behaves *ideally* (i.e it obeys the ideal gas law: $PV = nRT$, where P is the pressure; V is the volume; n is the number of kmoles; T is the temperature in K; and R is the ideal gas constant), the partial pressure of each component in a mixture volume V is given by: $p_A V = n_A RT$, etc. Furthermore, the partial pressure of each component is also equal to the product of the mole fraction (y_A) and total pressure, i.e. $p_A = y_A P$ etc.

1.2.3 Composition of Air-Water Mixtures

Air is extensively used in food processing. Its *humidity* is normally expressed on a dry basis; it is known as *absolute humidity (Y)* and defined as:

$$Y = \frac{mass\ of\ water\ vapour\ (kg)}{mass\ of\ dry\ air\ (kg)} \tag{1.6}$$

At any given temperature, there is a maximum amount of water vapour which can be associated with a given mass of dry air; the humidity of air under this condition is known as *saturation humidity* (Y_0). The ratio of absolute humidity to saturation humidity is known as *relative humidity* (RH), and it is often expressed as a percentage, i.e. RH $= 100Y/Y_0$. The water vapour associated with air exerts its own partial pressure, and when the air is saturated, the partial pressure is known as the *vapour pressure*. It follows that vapour pressure is, normally, only dependent on temperature. It increases with temperature and approaches the total pressure. When the vapour pressure is equal to the total pressure, the liquid is said to boil.

In general, when a free liquid surface is formed, exchange of molecules occurs across the interface in either direction, i.e. molecules evaporate from the liquid and condense back into it. Initially, there is net evaporation, but after a while, evaporation and condensation rates become equal and the space immediately above the interface becomes saturated. This phenomenon can be used to determine the extent to which a gas phase is saturated with any vapour. Indeed, this method is commonly used to determine the humidity of air. A thermometer, with a wet cloth wrapped around its bulb, shows a lower temperature than ambient, due to the evaporation of water from the bulb and the consequent loss of latent heat from it; this temperature is known as the *wet-bulb temperature*. The difference between ambient and wet-bulb temperature can be used as an indicator of the extent of saturation of air. If the two temperatures are very different, a substantial amount of evaporation has taken place, indicating that the air was not highly saturated in the first place. If the two temperatures are virtually the same, then evaporation has been negligible and therefore the air is nearly saturated. It must be pointed out that water evaporation from the thermometer bulb has been assumed to be *adiabatic*, i.e. a process which did not require the transfer of heat to or from the surroundings.

In the context of air-water mixtures, there is yet another temperature which can indicate humidity – *the dew point temperature*. This is the lowest temperature to which the mixture can be cooled without causing condensation. Once the temperature goes below this value, moisture begins to condense and form "dew". In other words, the mixture is fully saturated at the dew point. Thus, if there are two different air-water mixtures at any given temperature, their dew points will be different and the one with the lower dew point has a lower humidity.

Absolute humidity, relative humidity, vapour pressure, and wet-bulb and dew point temperatures are key properties of air-water mixtures. These, together with several other properties, are compiled in a chart known as the psychrometric chart, which, amongst other properties, can be used to estimate humidity from the results of dry and wet-bulb temperatures, or dew point measurements; see Fig. 1.1.

Problem 1.1 *Air enters a food dryer at 85 °C and a relative humidity of 2%. During adiabatic drying, it is cooled to a temperature that is 10 °C above its dew point. What is the condition of the exhaust air leaving the dryer?*

Air at 85 °C and 2% RH lies (approximately) on the 305 K adiabatic saturation line (the 2% RH curve is not explicitly shown in the Fig. 1.1). During drying, it follows this adiabatic saturation line, but leaves the dryer at a temperature that is 10 °C above the dew point. Given this constraint, a trial and error calculation is necessary. Let us assume that the air leaves at 80% RH. From the chart (Fig. 1.1), the outlet humidity is approximately 0.035 kg water/kg dry air; the outlet temperature is 38 °C; and the dew point is 33 °C, giving a temperature difference of only 5 °C. It therefore appears that a lower relative humidity of the exhaust air may have to be

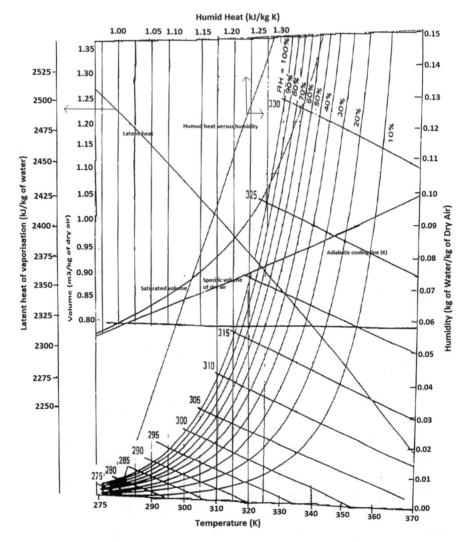

Fig. 1.1 Psychrometric chart for determining the properties of humid air. (Reproduced with permission from Carrier)

assumed. If RH is taken to be 60%, the outlet humidity is approximately 0.031 kg water/kg dry air; the outlet temperature is 42 °C; and the dew point is 32 °C, giving a temperature difference of 10 °C. It may be noted that, normally, a minimum difference of 10 °C is maintained between the exhaust air temperature and its dew point in order to prevent condensation, especially in the downstream gas cleaning equipment.

Fig. 1.2 Mass balance over
a control volume

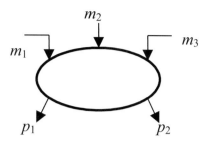

1.3   Mass Balance

1.3.1   Principles

Mass balance around individual equipment or around a process plant is based on the
law of conservation of mass which simply states that matter can neither be created
nor destroyed. This implies that mass or material entering a system (which may be an
equipment or a whole process) must be fully accounted for, and this is normally done
by writing equations which close the mass balance for individual components as well
as for the sum total of all components. In order to do so, the system's boundaries
must be defined, and the region enclosed by the boundaries is known as the *control
volume*. Consider the control volume defined in Fig. 1.2 and examine the inputs and
outputs. Two types of mass balances can be written around this control volume:
(i) mass balance over a period of time and (ii) mass balance at any given instant. If
m_1, m_2 and m_3 are the masses entering in kg and p_1 and p_2 are the masses of products
leaving over the same period of time, the mass conservation principle says that the
sum total of mass entering must be equal to the sum total of: mass leaving and mass
accumulated within this control volume. In other words,

$$m_1 + m_2 + m_3 = p_1 + p_2 + mass\ accumulated \tag{1.7}$$

If, on the other hand, the masses are continuously entering and leaving the control
volume, it would be more appropriate to write down an *instantaneous* mass balance
equation, which is essentially Eq. 1.7, except that m_1, m_2, m_3, p_1 and p_2 are *mass flow
rates* (kg s^{-1}), and the accumulation term essentially reflects the *rate of accumulation*. If the mass flow rates are all *uniform*, there will be no net accumulation within
the control volume, and the system is said to operate under *steady state*.

The *basis* of a mass balance equation must be clearly defined before any calculations are undertaken. The masses of all other components are then calculated
relative to this defined base. It is advisable to choose a basis that is most convenient
for calculations, and it is absolutely necessary not to change the basis during the
calculations. In many cases, there is no right or wrong choice – the same answer

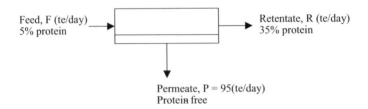

Fig. 1.3 Attached to Problem 1.2

should be obtained regardless of the basis. However, experience will show that a particular choice will speed up calculations. In many cases, a particular material enters the process in just one stream and leaves the process unchanged in one stream. For example, during drying, the moisture in air changes, but the components of dry air (mainly oxygen and Nitrogen), enter and leave the system unchanged. Such a component is known as *tie substance*, and if it can be identified, its use greatly simplifies and speeds up mass balance calculations.

The methods used in mass balances can be best illustrated by solving examples.

Problem 1.2 *A solution containing 5% proteins is concentrated to 35% in the retentate of a membrane separator, while the permeate is protein-free, as shown in the* Fig. 1.3. *If the separator produces 95 tonnes of permeate per day, what are the flows of the streams involved?*

Basis: 95 tonnes permeate per day. This basis has already been specified in the question. There are two components involved: water and protein; and two mass balance equations can be written, one for each component. However, we write down one *component mass balance* equation and an *overall mass balance* equation, giving two equations in all. Indeed, in an *n* component system, it is customary to write down any set of *(n − 1)* component mass balance equations and an overall mass balance equation, giving *n* equations in all.

In this case, overall mass balance: $F = P + R$ where $P = 95$ *te*; mass balance for proteins yields: $0.05F = 0.35R$. The two equations can be simultaneously solved to give: $F = 110.8$ te; $R = 15.8$ te

Problem 1.3 *In the manufacture of sugar from beet, the sugar (and other solubles) are extracted from cossettes by means of hot water in a counter-current diffuser. The residual pulp leaves at a moisture content of 90%. It then passes through a mechanical press in which the moisture content is reduced to 70%. Further moisture removal occurs in rotary dryers, before the dried pulp, at 10% moisture, is suitable for cattle feed. Calculate, per tonne of bone dry pulp solids: (a) the mass of water removed in the press and (b) the mass of water removed in the dryer.*

Basis: 1 tonne bone dry pulp entering. Once again, in this problem, the basis has been specified in the question. In any case, this is very suitable, since it can be treated as a tie substance.

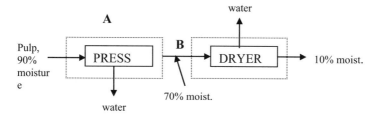

Fig. 1.4 Attached to Problem 1.3

(a) Since this part of the question concerns the Press, the *control volume* (A) has been constructed around it as shown in the Fig. 1.4. The pulp has two components - dry matter and moisture; hence two independent mass balance equations can be written down. One mass balance equation can account for moisture, while the other can reflect the *dry matter*

Dry matter balance: 1tonne enters the volume and all dry matter leaves with the pulp leaving the volume. If m_{p1} is the mass of wet pulp leaving the press, containing 30% moisture, $1 = 0.3\, m_{p1}$.

Moisture balance: Moisture entering per tonne of dry matter = 9 tonnes. A part of this water is pressed out, say x_p kg, while the remaining water finds its way with the pulp leaving the press. Thus $9 = x_p + 0.7\, m_{p1}$.

The two equations can be solved to yield $m_{p1} = 3.33$ te and $x_p = 6.67$ te. Thus 6.67 te of water is removed by the press for every tonne of dry pulp entering.

(b) Following the same procedure as above, two mass balances can be written down around *control volume* B. *Dry matter balance*: 1 tonne enters the volume and if m_{p2} kg pulp leaves the dryer containing 10% moisture, $1 = 0.9\, m_{p2}$. *Water balance*: Water entering the volume $= 0.7 \times 3.33$ te and water leaving with the pulp is $0.1\, m_{p2}$. If x_D is the mass of m_{p2}. water excluded in the dryer, $0.7 \times 3.33 = 0.1\, m_{p2} + x_D$. The two water balance equations can be solved to yield $m_{p2} = 1.11$ te and $x_D = 2.22$ te.

1.3.2 Recycles

Recycle streams are very commonly used in process technology, mainly to conserve material and to improve process efficiency. For example, a fraction of the air leaving a dryer can be recycled by mixing with the incoming air, in order to utilise more effectively its drying capacity as well as heat. In the operation of a distillation column, it is common practice to condense the distilled vapour and put a fraction of it back into the column as a recycle stream in order to enrich component separation.

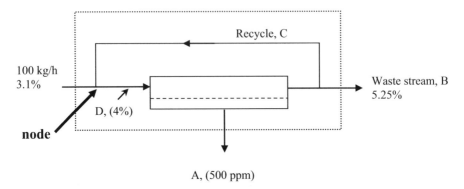

Fig. 1.5 Attached Problem 1.4

Problem 1.4 *A reverse osmosis plant is used to produce potable water from a feed stream containing 3.1% by weight of contaminant (Fig. 1.5). The purified stream is to contain 500 ppm of the contaminant. All other concentrations are indicated in Fig. 1.5. Calculate the flow rates and compositions of the streams A, B, C, and D.*

Basis: 100 kg feed. A control volume covering the whole process can be defined as shown in the figure, and (i) an overall and (ii) a component balance can be written on the basis of the streams entering and leaving this volume. Overall balance: $A + B = 100$; contaminant balance: $500 \times 10^{-6}A + 0.0525B = 3.1$, which can be solved simultaneously, to yield: $A = 41.35$ kg; and $B = 58.65$ kg.

Now given that the composition of stream D is known, the overall and component mass balances can be written over the *node point*: $100 + C = D$; $3.1 + 0.0525 C = 0.04 D$, which can be simultaneously solved to give: $C = 72.00$ kg and $D = 172.00$ kg. Finally, the compositions of the streams are:

Stream	Contaminant (kg)	Water (kg)
A	$41.35 \times 500 \times 10^{-6} = 0.02$	$41.35 - 0.02 = 41.33$
B	$58.65 \times 0.0525 = 3.08$	$58.65 - 3.08 = 55.57$
C	$72.00 \times 0.0525 = 3.78$	$72.00 - 3.78 = 68.22$
D	$172.00 \times 0.04 = 6.88$	$172.00 - 6.88 = 165.12$

1.3.3 Chemical Reactions

In the examples considered above, the chemical composition of the species has remained unchanged, although physical changes have occurred. When chemical changes are involved in any process, in addition to the closure of the overall mass balance, balances on individual elements must also close. Further, if inert species are involved, a balance on them must also close.

A knowledge of the stoichiometry of the reaction or reactions involved is therefore a prerequisite to write down elemental mass balance equations. This is fairly simple if the chemical formulae of the species, and the exact mechanisms of all the reactions involved are known. In most food and biochemical systems, the processes may be too complicated to be represented by one or few chemical equations. This is certainly true in the case of fermentations, where thousands of reactions are involved in cell and metabolite synthesis. An approach that is commonly taken, mainly due to its relative simplicity, is known as the black box model, where all cellular reactions are lumped into a single reaction for overall biomass growth. For instance, the cultivation of *Saccharomyces cerevisiae* under aerobic conditions where glucose is the carbon source and ammonia is the nitrogen source, can be represented as:

$$CH_{1.83}O_{0.56}N_{0.17} + Y_{xe}CH_3O_{0.5} + Y_{xc}CO_2 + Y_{xw}H_2O - Y_{xs}CH_2O - Y_{xo}O_2 - Y_{xN}NH_3 = 0$$

In such equations, the terms with positive stoichiometric factors are the products and those with negative stoichiometric factors are the reactants; the factors themselves are known as yield coefficients. For example, Y_{xe} is the yield coefficient of ethanol and Y_{xs} is the yield coefficient of glucose. Further, it is also customary to use C-mole basis (i.e. CH_2O represents glucose and $CH_3O_{0.5}$ represents ethanol. The elemental composition of the microorganism depends on the macromolecular content. It is also necessary to note that formula of a given microorganism is not fixed, but it depends on the conditions under which it is grown. Although the above equation is not mechanistic, it must satisfy all consistency requirements. In other words, the number of atoms of each element must be balanced, i.e.:

C balance: $1 + Y_{xe} + Y_{xe} + Y_{xc} - Y_{xs} = 0$
H balance: $1.83 + 3Y_{xe} + 2Y_{xw} - 2Y_{xs} - 3Y_{xN} = 0$
balance: $0.56 + 0.5Y_{xe} + 2Y_{xc} + Y_{xw} - Y_{xs} - 2Y_{xo} = 0$
N balance: $0.17 - Y_{xN} = 0$

The yield coefficients for systems containing biomass are determined experimentally and are therefore susceptible to experimental errors. Elemental mass balance equations, such as those written above, can be used to check the experimental results.

Problem 1.5 *100 kg of propane (C_3H_8) is burned with a stoichiometric quantity of air (21 vol % oxygen and the rest nitrogen). Assuming complete combustion, calculate the composition of the products on a mass basis.*

Basis: 100 kg of propane
Equation: $C_3H_8 + 5O_2 \rightarrow 3CO_2 + 4H_2O$

Since all the propane is combusted, the product mixture will consist of CO_2, H_2O and nitrogen which is present in air and remains unreacted. Further, since the stoichiometry indicates the molar proportion by which the various species are formed or consumed, it is necessary to work in terms of moles.

Reactants: Propane $= 100$ kg $= 100/44 = 2.27$ kmol, which requires 5×2.27 kmol of oxygen for complete combustion $= 11.36$ kmol. Given that air contains nitrogen to oxygen in the molar ratio 79/21 (i.e. the same as volume ratio), 11.36 kmol of oxygen will be associated with 42.75 kmol nitrogen (or $42.75 \times 28 = 1197$ kg); this will find its way in the product. From the stoichiometry of the reaction, the mass of carbon dioxide formed is $3 \times 2.27 = 6.81$ kmol (or $6.81 \times 44 = 299.64$ kg); and that of water vapour fomed is $4 \times 2.27 = 9.08$ kmol (or $9.08 \times 18 = 163.44$ kg). Thus the product mixture contains: 1197 kg nitrogen, 299.64 kg carbon dioxide and 163.44 kg water vapour, giving a total mass of 1660.08 kg. It would be instructive to check if this figure tallys with the total mass of the reactants.

Problem 1.6 *When yeast cells are grown using methane (CH_4) as the carbon substrate, 50 wt% of the substrate is converted to cells. The process stoichiometry can be represented as:*

$$CH_4 + a_1O_2 + a_2NH_3 \rightarrow b_1C_4H_7O_{1.65}N_{0.93} + b_2CO_2 + b_3H_2O$$

Calculate:

(a) *The stoichiometric coefficients: a_1, a_2, b_1, b_2 and b_3.*
(b) *The yield coefficient on oxygen, Y_{O2} (g cell per g oxygen consumed)*
(c) *The respiratory coefficient, RQ (moles CO_2 per mole O_2)*

$$CH_4 + a_1O_2 + a_2NH_3 \rightarrow b_1C_4H_7O_{1.65}\,N_{0.93} + b_2CO_2 + b_3H_2O$$

C-balance yields:	1	$=$	$4b_1 + b_2$
H balance:	$4 + 3a_2$	$=$	$7b_1 + 2b_3$
O balance:	$2a_1$	$=$	$1.65b_1 + 2b_2 + b_3$
N balance:	a_2	$=$	$0.93b_1$

Since 50% by wt of substrate is converted into cells, we have: $16/2 = b_1$ (94.42); hence $b_1 = 0.085$, which gives $b_2 = 0.661$; $a_2 = 0.079$; $b_3 = 1.821$; and $a_1 = 1.641$

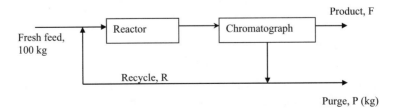

Fresh feed, 100 kg — Reactor → Chromatograph → Product, F

Recycle, R

Purge, P (kg)

Fig. 1.6 Attached to Problem 1.7

b. $Y_{O2} = (94.42 \times 0.085) / (32 \times 1.641) = 0.153$

c. moles CO_2 /mole $O_2 = 0.661/1.641 = 0.403$

Problem 1.7 *High fructose syrup is produced by enzyme catalysed isomerisation of glucose, which is a reversible reaction having an equilibrium constant of one. In order to achieve a higher conversion of glucose, the reaction products are passed through a chromatograph, which achieves a complete separation of glucose and fructose. The glucose stream is then recycled to the reactor as shown in* Fig. 1.6 *In practice, the fresh stream contains 98% glucose and 2% oligosaccharides (i.e. carbohydrates containing 2-8 monosaccharide units); the latter does not take part in the reaction and separates with the glucose in the chromatograph. Assuming that the reaction products are at equilibrium, calculate the recycle rate, R, and the fractional conversion of glucose, f, if the percentage of oligosaccharides in the purge stream is set to 10%. It may be noted that without the purge stream, the percentage of oligosaccharides will progressively build up.*

Basis: 100 kg of fresh feed containing 98 kg glucose and 2 kg oligosaccharide. The overall mass balance equation can be written as: $F + P = 100$. Since the percentage of oligosaccharide in the purge stream is set to 10%, a mass balance of oligosaccharide yields: $0.1P = 2$ or $P = 20$ kg; hence, $F = 80$ kg.

Note that the purge stream (and the recycle stream R) contains 90% glucose; and the amounts of glucose and fructose in the stream leaving the reactor are equal since the reaction products are in equilibriim and the equilibrium constant is 1.

Mass balance over the reactor:

	Inflow	Outflow
Glucose	$98 + 0.9R$	$49 + 0.45R$
Fructose	0	$49 + 0.45R$
Inert (Oligosaccharide)	$2 + 0.1R$	$2 + 0.1R$

All the fructose leaving the reactor ends up as the product stream F, i.e. $49 + 0.45R = 80$, giving $R = 68.9$ kg; and the fractional coversion of glucose into fructose is: $80/98 = 0.816$.

In the above problem, it is instructive to set up a spread sheet which relates the percentage of inerts in the purge, the recycle R and fractional conversion, f; and discuss the existence of an optimum recycle ratio at which production costs can be minimised.

1.4 Thermal Energy or Enthalpy

1.4.1 Principles

The two most common forms of energy involved in food processing are mechanical and thermal, although these forms of energy are ultimately obtained from electrical energy that should be available on site. The SI unit of energy is *Joule* (J). A Joule is too small a quantity to have any practical significance; hence kJ or MJ are the common units used in practice. Further, the rate of energy transferred, i.e. *power*, are encountered in common practice, and its units is kJ s^{-1} or kW. There are three common forms of mechanical energy used in process equipment: potential (mgh), kinetic ($1/2\ mv^2$) and pressure (PV); these forms will be discussed further in this chapter. In this section, the thermal energy, and balances involving it, will be considered in detail.

There are two types of thermal energy. One relates to the heat added (or removed) from a substance in order to raise (or lower) its temperature; this heat is known as *sensible heat*. In general, the heat needed to change the temperature of 1 kg a substance by 1 Kelvin (K) is known as its *specific heat* (C_p) with units kJ kg^{-1} K^{-1}; and the sensible heat needed to change the temperature of m kg of the substance by ΔT is: $mC_p\ \Delta T$. The second form of thermal energy is the *latent heat*, which is defined as the heat necessary to change the physical state of 1 kg of any substance, i.e. solid to liquid or liquid to vapour, or *vice versa*. The net latent heat for changing the state of m kg is therefore given by mL where L is the latent heat expressed in, say, kJ kg^{-1}. If heat transfer rates (expressed in, say, kW) are being considered in a situation involving materials which are flowing through the process, m will represent the mass flow rate (kg s^{-1}).

It is apparent from the above discussion that the thermal energy level of any substance is not an absolute property; it has to be defined relative to a base temperature. If T_0 is the base temperature and the substance is in a solid state at this temperature, its enthalpy at any other temperature T, say, in the vapour state, is obtained by calculating the energy needed to heat 1 kg of substance from T_0 to T, going through melting at T_f and boiling at T_b. Thus, the change in *enthalpy* (ΔH) is given by:

$$\Delta H = C_{pS}\left(T_f - T_0\right) + L_f + C_{PL}(T_b - T_f) + L_b + C_{PV}(T_b - T) \tag{1.8}$$

where C_{pS}, C_{pL}, and C_{pV} are the specific heats of the solid, liquid and vapour, and L_f and L_b represent latent heats of fusion and vapourisation at the respective temperatures. It may be noted that the specific heat has been assumed to be independent of temperature, which is often not the case. If C_p is a function of temperature, Eq. (1.8) has to be modified and the change in specific enthalpy, ΔH, is given by:

$$\Delta H = \int_{T_0}^{T_f} C_{pS}dT + L_f + \int_{T_f}^{T_b} C_{pL}dT + L_b + \int_{T_0}^{T} C_{pV}dT \tag{1.9}$$

The variation of C_p as a function of temperature is often expressed as a polynomial, e.g. $C_p = a + bT + cT^2$, where a, b, and c are constants depending on the material under consideration.

Steam is commonly used as a heating medium in food and bioprocessing because:
(1) it can be generated in a manner that renders it clean enough for injection into food

products; (2) the temperatures at which it can be generated are suitable for sterilising and promoting desired changes in food products; (3) it has a very high enthalpy, especially latent heat, which can be easily transferred at constant temperature; (4) it can easily be recirculated, allowing re-use of condensate; and (5) its temperature can be accurately controlled by controlling the pressure. The thermal properties of steam have been widely studied and published in the form of *steam tables*. The steam table, shown in Appendix 1.1, lists the enthalpy of water and its vapour at different pressures and temperatures. The difference between the enthalpies of solid and liquid at any given temperature and pressure is the latent heat of fusion at that temperature; likewise, the difference between the enthalpies of vapour and liquid gives the latent heat of evaporation. It may also be noted that a vapour at the boiling point is said to be *saturated*, and if its temperature is raised above this value, it is said to be *superheated*.

Example 1.1

What is the enthalpy of saturated steam at 430 K based on water at 300 K. Assume the following constant physical properties: specific heat of liquid water = 4.18 kJkg^{-1} K^{-1}; specific heat of steam = 2.00 kJkg^{-1} K^{-1}; and latent heat of vapourisation at 373 K = 2258 kJkg^{-1}.

In order to calculate the enthalpy of steam at 430 K we have to account for the following enthalpy changes: heating water from 300 K to its boiling point 373 K, vapourisation at 373 K, and heating the vapour from 373 K to 430 K. Therefore, the change in enthalpy is:

$$\Delta H = 4.18(373 - 300) + 2258 + 2.00(430 - 373) = 2677 \text{ kJ/kg.}$$

Example 1.2

What is the enthalpy of air at 310 K, 50% RH, if the heat capacities (i.e. specific heats) of air and water are 1.00 and 2.00 kJ/kg and the latent heat of vapourisation of water at 273 K is 2500 kJ/kg.

In this problem the base state has not been specified; an assumption must be made in this respect. It is common practice to base the enthalpy of air-water mixture on liquid water at 273 K.

Using psychrometric chart (Fig. 1.1), the absolute humidity of air is 0.038 kg moisture/kg dry air. In order to obtain 1 kg of moist air at 310 K and 50% RH, we start with 0.037 kg (i.e. 0.038/1.038) water and 0.963 kg of dry air, both at 273 K. The water is evaporated at 273 K and the resulting vapour is heated to 310 K. The enthalpy change for this process is: 0.037 × 2500 + 0.037 × 2.00 × (310 − 273) = 95.24 kJ. The air is then heated from 273 K to 310 K; and the enthalpy change is: 0.963 × 1.00 × (310 − 273) = 35.63 kJ. The net enthalpy of 1 kg of moist air is therefore: 95.24 + 35.63 = 130.87 kJ.

Example 1.3

1 m³ of orange juice (density = 1020 kg m⁻³), initially at 295 K, is heated for 15 minutes in an unlagged batch tank, using an electric heater rated 100 kW. For each of the cases mentioned below, estimate the final temperature if the average heat loss to the surroundings is 5 kW: Case 1 – neglect evaporation losses from the tank; Case 2 – assume that 1% of the starting mass is lost by evaporation. The mean heat capacity of the juice is 4.2 kJkg⁻¹ K⁻¹; and the mean latent heat of vapourisation between 295 and 320 K can be assumed to be 2425 kJkg⁻¹.

Basis: $1m^3$ or 1020 kg of juice.

Energy supplied by heater over 15 minutes $= 100 \times 15 \times 60 = 90\,000$ kJ, of which $5 \times 15 \times 60 = 4500$ kJ is lost to the surroundings. The net energy absorbed by the tank contents is therefore 85500 kJ.

Case 1: The above energy is utilised fully for raising the temperature of the juice. Hence, $85500 = mC_p(T - T_i)$ where T is the final temperature and T_i is the initial temperature. Given that $m = 1020$ kg, $C_p = 4.2$ kJkg⁻¹ K⁻¹ and $T_i = 295$ K, we have $T = 314.96$ K.

Case 2: The net energy absorbed by the tank contents is partly utilised for evaporation, while the remaining heats the juice. Given that 1% of the initial mass evaporates, the energy used up in this process is: $10.2 \times 2425 = 24735$ kJ, leaving $(85500 - 24735)$ or 60765 kJ for raising the juice temperature. Employing the same method as in case 1, except, replacing the energy value by the new figure and using an average mass of $(1020 + 0.9 \times 1020)/2 = 969$ kg, we have $T = 309.9$ K.

It is necessary to note that a number of assumptions have been made in the above calculations, some of which have not been explicitly stated. It is useful to list all assumptions and, if possible, develop a more rigorous calculation procedure.

1.4.2   Enthalpy of Reactions

Most chemical and biochemical reactions evolve or absorb heat, which must be taken into account while performing energy balances. Whether heat is evolved or absorbed in a reaction depends on the *standard Heat of formation* of the reactants and products. *The Standard Heat of formation of a compound is the change in enthalpy when 1 kmol of a compound is formed from its elements in stoichiometric proportions, beginning and ending at 25 °C and 1 bar, with specified states of elements and compounds.* The standard heat of reaction is defined similarly, and it represents the enthalpy change resulting from a reaction beginning and ending at 25 °C, and 1 bar, with specified states of elements and compounds. The standard heat of reaction can be calculated from the standard heats bof formation of the reactants an products through Hess' law, which for the reaction:

$xA + yB \rightarrow zC$ is calculated as:

$$\Delta H_R^0 = z\Delta H_{fC}^0 - \left(x\Delta H_{RA}^0 + y\Delta H_{RB}^0\right) \tag{1.10}$$

Problem 1.8 *One of the fundamental reactions in biology is the combustion of glucose in living systems to produce energy. The reaction can be represented as:*

$$C_6H_{12}O_6 + 6O_2 \rightarrow 6CO_2 + 6H_2O + \Delta H$$

What is the heat released per kg glucose using the following values for the heats of formation in MJ kmol^{-1} (note that negative values indicate that heat is evolved): Glucose = −1274; Oxygen = 0 (it may be noted that the standard heat of formation of pure elements is zero); Carbon dioxide = −414; and water = −286.

From Eq. 1.10, $\Delta H = 6 \times (-414) + 6 \times (-286) - (-1274) - 0 = -2926$ MJ/ kmol glucose, which is $-2929/180 = -16.3$ kJ/kg.

Problem 1.9 *Propane is burned with 50% excess air to ensure complete combustion. Assuming that the fuel and air enter the burner at 18C and that there are no combustion losses, estimate the temperature of the final product. Net heating value of Propane = 2044 MJ/kmol; and heat capacities are: oxygen = 1.04 kJ/kg/K; nitrogen; 1.13 kJ/kg/K; carbon dioxide = 1.21 kJ/kg/K; and water vapour = 2.09 kJ/kg/K. Assume that the reference temperature is 18C, gaseous state.*

Basis: 1 kmol of propane
The combustion equation is: $C_3H_8 + 5O_2 \rightarrow 3CO_2 + 4H_2O$

Since 1 kmol of propane needs 5 kmol oxygen and air contains 21 mole% of oxygen, the amount of air needed for combustion $= 5 \times 100/21 = 23.81$ kmol. Given that 50% excess air is used, the moles of air $= 23.81 \times 1.5 = 35.7$ kmol. This air will contain 7.5 kmol (or 240 kg) oxygen and 28.2 kmol (or 790 kg nitrogen). Of this $7.5 - 5 = 2.5$ kmol of oxygen (or 80 kg) passes unreacted, as does all nitrogen. The exit gas will also contain 3 kmol (or 132 kg) of carbon dioxide

Since both the reactants are in the gaseous state at 18C, their enthalpies are zero. The heat of combustion released is 2044 MJ/kmol propane. The total heat content of products is as follows:

Oxygen: $80 \times 1.04 \times (T\text{-}18)$; Nitrogen: $790 \times 1.13 \times (T\text{-}18)$; Carbon dioxide: $132 \times 1.21 \times (T\text{-}18)$; and water vapour $= 72 \times 2.09 \times (T\text{-}18)$. Since the heat evolved by reaction should be associated with the products, the sum total of these enthalpies must be equal to 2.044×10^6 kJ, whence T = 1607 C.

1.5 Enthalpy Balances

1.5.1 Principles

The law of conservation of energy forms the basis of enthalpy balance, and the *First Law of Thermodynamics* is widely used as a means of stating the conservation principle. According to this law, the external heat supplied to system (Q) is partly utilised to change the internal energy (ΔE), which is enthalpy, for all practical purposes) and the remaining fraction is utilised for doing mechanical work (W). In other words,

$$Q = \Delta E + W \tag{1.11}$$

If work is done *on* the system, instead of the *system doing work*, W is assumed to be negative. Equation (1.11) also assumes that no mass flows into or out of the system; i.e the system is *closed*. If, on the other hand, an open system is considered, energy will flow in and out of it through the masses entering and leaving; and Eq. (1.11) can be modified to:

$$Q + m_i e_i = \Delta E + W + m_o e_o \tag{1.12}$$

where m_i and m_o are the masses entering and leaving the system; e_i and e_o are the energy levels per unit mass of the streams entering and leaving the system. It may also be noted that in this case Q is the *rate* at which heat enters the system; W is the rate at which it does work; and ΔE is the rate at which the internal energy changes. All terms are expressed in units of power.

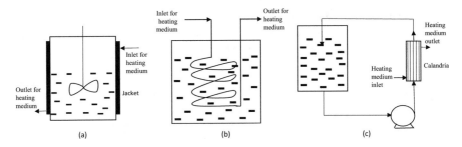

Fig. 1.7 Methods of heating: (**a**) jacket fitted to the vessel, (**b**) immersed coil (**c**) use of an external heat exchanger or calandria

1.5.2   Enthalpy Balances in Process Equipment

Heat can either be added to (or removed from) a process fluid by bringing it into contact with a heating (or cooling) medium which is normally a fluid. The contact between the two fluids is normally *indirect*, i.e. the two fluids exchange energy across solid walls without dispersing into one another. In practical terms, if a process fluid is to be heated in a vessel, the heat may be transferred either by passing the hot fluid through a jacket fitted to the vessel (Fig. 1.7a) or through a coil immersed in the fluid (Fig. 1.7b). Alternatively, the fluid can be heated externally by passing through a heat exchanger (Fig. 1.7c). In all these cases, thermal energy balance can be closed by assuming that the heat lost by the heating medium is equal to the heat gained by the process fluid. Of course, practical heating operations can involve evaporation losses as well as energy lost to the surroundings, which, if significant, must be taken into account in order to close the thermal energy balance (see Problem 1.8). Normally, such losses are negligible in well insulated heat exchangers.

Problem 1.10   *400 kg per hour of sauce (specific heat = 3.95 kJ kg^{-1} K^{-1}) is to be heated from 60 °C to 115 °C in a heat exchanger by steam which has a quality of 90% and is available at 125 °C. Determine the steam flow rate required assuming that the condensate does not subcool in the heat exchanger (i.e. the steam and condensate are both at 125 °C).*

The energy required to heat the sauce $= 400 \times 3.95 \times (115 - 60) = 86900$ kJ h^{-1}. Given that the steam has a quality of 90%, the enthalpy of this wet steam can be determined from steam tables (Appendix 1.1) as: $H_1 = 0.9 \times 2713 + 0.1 \times 524 = 2494$ kJ kg^{-1}. Since there is no subcooling, the enthalpy of condensate at 125 °C $H_2 = 524$ kJ kg^{-1}. Therefore the heat lost by the steam is $(2494 - 524) = 1970$ kJ kg^{-1}. If m_s kg h^{-1} is the mass of steam required, heat lost by the steam $= m_s \times 1970$, which should be equal to the heat absorbed by the sauce (assuming no heat losses). In other words, $1970\, m_s = 86900$; or $m_s = 44.1$ kg h^{-1}.

Problem 1.11   *15000 kg h^{-1} of milk at 15 °C is sterilised by heating to 135 °C in a heat exchanger by using saturated steam available at 175 °C. It is then held at this temperature for 7 s and cooled in another heat exchanger to 35 °C using water at 18 °C. If the steam does not undergo any subcooling in the heater and the cooling water leaves the cooler at 25 °C, calculate: (i) the rate of heat transfer in the heater, (ii) the flow rate of steam required, (iii) the rate of heat transfer in the cooler, and (iv) the flow rate of cooling water required. Assume that milk and water have the same thermal properties (eg specific heat = 4.2 kJkg^{-1} K^{-1}).*

Basis: 15000 kg h^{-1} or $15000/3600 = 4.17$ kg s^{-1} of milk.

Heat balance in the heater simply states that the rate at which milk gains heat = the rate at which steam loses heat. The rate at which milk gains heat $= 4.17 \times 4.2 \times (135 - 15) = 2101.68$ kW, which is the rate of heat transfer in the heater. If m_s is the mass flow rate of steam expressed in kg s^{-1} and 2122 kJ kg^{-1} is its latent heat at 175 °C (referring to steam tables), $2122\, m_s = 2101.68$, which yields $m_s = 0.99$ kg s^{-1}.

Following the same procedure, the heat lost by milk in the cooler = heat gained by the cooling water. The heat lost by milk = $4.16 \times 4.2 \times (175 - 35) = 2446.1$ kW which is the rate of heat transfer in the cooler. If m_w kg s^{-1} is the mass flow rate of cooling water entering the cooler, the heat gained = $m_w \times 4.2 \times (25 - 18) = 29.4 \, m_w$ kW, which, when equated to 2446.1, gives $m_w = 83.2$ kg s^{-1}.

It may be noted that this is not a very energy efficient way of running a steriliser. Given that the milk is first of all heated to a very high temperature and then rapidly cooled down, the heat associated with the hot milk itself can be used to preheat the incoming cold milk. This is indeed the normal practice.

Problem 1.12 *Apple juice (1000 kg) contains 12% total solids. Using the data provided below estimate how much energy would be required to freeze this juice from an initial temperature of 20 °C down to −18 °C.*

Data

Specific heat of water = 4.18 kJ kg^{-1} K^{-1}
Specific heat of apple juice solids = 1.90 kJ kg^{-1} K^{-1}
Latent heat of fusion for ice = 330 kJ kg^{-1}
Specific heat of frozen apple juice = 2.00 kJ kg^{-1} K^{-1}

Basis: 1000 kg Apple juice, which contains 120 kg solids and 880 kg water, initially at 20 °C .

An average specific heat of liquid apple juice can be estimated a weighted average of solid and water content, i.e. $0.12 \times 1.90 + 0.88 \times 4.18 = 3.91$ kJ kg^{-1} K^{-1}. The energy required to cool 100 kg of apple juice from 20 °C to 0 °C (assuming the juice freezing point to be 0 °C) is $1000 \times 3.91 \times (20 - 0) = 78200$ kJ. The latent heat to be removed in order to freeze the 880 kg of water present in the juice is: $880 \times 330 = 290400$ kJ. Finally, the sensible heat removed to cool the frozen apple juice from 0 °C to –18 °C is: $1000 \times 2 \times 18 = 36000$ kJ. The total energy required is therefore: 404600 kJ

Appendix 1.1: Properties of Saturated Steam (Data Taken From https://www.ohio.edu/mechanical/thermo/property_tables/H2O/H2O_TempSat1.html; and www.efunda.com); Subscripts f -Liquid Phase, g – Vapour Phase and fg – Phase Change

Temp (°C)	Pressure (MPa)	Specific Volume (m^3 kg^{-1})		Enthalpy (kJ kg^{-1})			Entropy (kJ kg^{-1}.K^{-1})		
T_{sat}	P_{sat}	v_f	v_g	h_f	h_{fg}	h_g	S_f	s_{fg}	s_g
0.01	0.00061	0.00100	205.99	0.001	2500.9	2500.9	0	9.1555	9.1555
5	0.00087	0.00100	147.01	21.0	2489.1	2510.1	0.0763	8.9485	9.0248
10	0.00123	0.00100	106.30	42.0	2477.2	2519.2	0.1511	8.7487	8.8998

(continued)

Temp (°C)	Pressure (MPa)	Specific Volume (m³ kg⁻¹)		Enthalpy (kJ kg⁻¹)			Entropy (kJ kg⁻¹.K⁻¹)		
T_{sat}	P_{sat}	v_f	v_g	h_f	h_{fg}	h_g	S_f	s_{fg}	s_g
15	0.00171	0.00100	77.875	63.0	2465.3	2528.3	0.2245	8.5558	8.7803
20	0.00234	0.00100	57.757	83.9	2453.5	2537.4	0.2965	8.3695	8.6660
25	0.00317	0.00100	43.337	104.8	2441.7	2546.5	0.3672	8.1894	8.5566
30	0.00425	0.00100	32.878	125.7	2429.8	2555.5	0.4368	8.0152	8.4520
35	0.00563	0.00101	25.205	146.6	2417.9	2564.5	0.5051	7.8466	8.3517
40	0.00739	0.00101	19.515	167.5	2406.0	2573.5	0.5724	7.6831	8.2555
45	0.00960	0.00101	15.252	188.4	2394.0	2582.4	0.6386	7.5247	8.1633
50	0.01235	0.00101	12.027	209.3	2382.0	2591.3	0.7038	7.3710	8.0748
55	0.01576	0.00102	9.5643	230.3	2369.8	2600.1	0.7680	7.2218	7.9898
60	0.01995	0.00102	7.6672	251.2	2357.6	2608.8	0.8313	7.0768	7.9081
65	0.02504	0.00102	6.1935	272.1	2345.4	2617.5	0.8937	6.9359	7.8296
70	0.03120	0.00102	5.0395	293.2	2333.0	2626.1	0.9551	6.7989	7.7540
75	0.03860	0.00103	4.1289	314.0	2320.6	2634.6	1.0158	6.6654	7.6812
80	0.04741	0.00103	3.4052	335.0	2308.0	2643.0	1.0756	6.5355	7.6111
85	0.05787	0.00103	2.8258	356.0	2295.3	2651.3	1.1346	6.4088	7.5434
90	0.07018	0.00104	2.3591	377.0	2282.5	2659.5	1.1929	6.2852	7.4781
95	0.08461	0.00104	1.9806	398.1	2269.5	2667.6	1.2504	6.1647	7.4151
100	0.10142	0.00104	1.6718	419.2	2256.4	2675.6	1.3072	6.0469	7.3541
110	0.14338	0.00105	1.2093	461.4	2229.7	2691.1	1.4188	5.8193	7.2381
120	0.19867	0.00106	0.8912	503.8	2202.1	2705.9	1.5279	5.6012	7.1291
130	0.27028	0.00107	0.66800	546.4	2173.7	2720.1	1.6346	5.3918	7.0264
140	0.36154	0.00108	0.50845	589.2	2144.2	2733.4	1.7392	5.1901	6.9293
150	0.47616	0.00109	0.39245	632.2	2113.7	2745.9	1.8418	4.9953	6.8371
160	0.61823	0.00110	0.30678	675.47	2082.0	2757.4	1.9426	4.8026	6.7491
170	0.79219	0.00111	0.24259	719.08	2048.8	2767.9	2.0417	4.6233	6.6650
180	1.0028	0.00112	0.19384	763.05	2014.2	2777.2	2.1392	4.4448	6.5840

References

Niranjan K (1994) Chemical engineering principles and food processing. Trends Food Sci Technol 5:20–23

Bibliography for Further Reading

Yanniotis S (2007) Solving problems in food engineering. Springer, pp 11–32
Doran PM (2012) Bioprocess engineering principles, 2nd edn. Elsevier, pp 87–176

Chapter 2
Elements of Fluid Flow

Keshavan Niranjan

Aims Given that most materials encountered in food processes are in the fluid state, this chapter aims to deal with the basic principles of fluid flow. In particular, the chapter considers how skin and form friction occurring at fluid-solid boundaries can be characterised and mitigated. The chapter covers flow of fluids through pipes and fittings, flow of fluids around particles for Newtonian and selected non Newtonian flows. In addition, the chapter also introduces flow through a bed of particles and fluidisation.

2.1 Introduction

Fluids are normally taken to mean gases and liquids, and they differ from the other normal state of matter – solid - in the sense that the latter has a definite shape while the former assume the shape of the containing vessel. In more rigorous terms, the distinguishing feature of a fluid is taken to be its inability to resist applied *shear forces*. Although a common sense based distinction can generally be drawn between solids and fluids, there are many food materials which are "borderline" cases because they exhibit properties appropriate to fluids as well as solids depending on the way they are examined, or on the timescale over which they are examined. For example, it is possible to associate "structure" with foods like yoghurt, which can also be made to "flow" under certain conditions. This chapter will essentially consider the flow properties of fluids, and discuss practical issues relating to the transport of fluids especially within food processing plants.

K. Niranjan
Department of Food and Nutritional Sciences, University of Reading, Reading, Berkshire, UK
e-mail: afsniran@reading.ac.uk

© Springer Nature Switzerland AG 2022
K. Niranjan, *Engineering Principles for Food Process and Product Realization*,
Food Engineering Series, https://doi.org/10.1007/978-3-031-07570-4_2

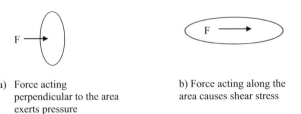

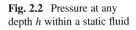

a) Force acting   b) Force acting along the
   perpendicular to the area   area causes shear stress
   exerts pressure

Fig. 2.1 Conceptualising (**a**) Pressure and (**b**) Shear stress. Both these parameters have the same
unit (Pa). (**a**) Force acting perpendicular to the area exerts pressure. (**b**) Force acting along the area
causes shear stress

Fig. 2.2 Pressure at any
depth h within a static fluid

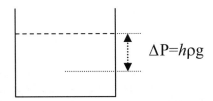

$$\Delta P = h\rho g$$

We start the discussion with two basic concepts: *pressure* and *shear stress*. Both
these terms refer to the force acting per unit area around any point in a fluid. The
distinguishing feature is in the orientation of the area relative to the force: pressure
refers to the force acting per unit area *normal to the force*, whereas, shear stress refers
to the force acting per unit area *that includes the line of action of the force*; see
Fig. 2.1. Both these properties share a unit: Nm^{-2} or Pa. It may be noted that
atmospheric pressure is approximately equal to 101 kPa.

Pressure is a property inherent within a fluid and it is exerted at every point,
regardless of whether a fluid is at rest or in motion. In a static fluid, the pressure at
any depth, measured in excess of that at its surface, is equal to $h\rho g$, where h is the
depth of the point under consideration; ρ is the fluid density; and g is the acceleration
due to gravity (Fig. 2.2).

Shear stress, on the other hand, causes a fluid to *flow*. Indeed, fluids can be
defined as materials which undergo continuous deformation when shear stress is
applied. If materials offer *no resistance* to applied shear, they are said to flow *ideally*,
and the theory of fluid mechanics evolved under the assumption that flows were
ideal. In practice, all fluids offer resistance, albeit to different extents, and *ideal
theories* have been extended to cover *real flows*.

Two basic terms relating to fluid flow (ideal or otherwise) will now be defined.
Flow rate is the amount of fluid flowing per unit time, and it can either be expressed
as the *volume* flowing per unit time (Q) or *mass* flowing per unit time ($\dot{m}$). The
volumetric flow rate has the units of m^3s^{-1}; and the mass flow rate is expressed in
kgs^{-1}. The two flow rates are obviously related through the fluid density: $Q = \frac{\dot{m}}{\rho}$
where ρ is the local fluid density at the point being considered. The second term is
the *superficial fluid velocity*, u, which is defined as the volumetric flow rate per unit
cross sectional area perpendicular to the flow; i.e. $u = Q/A$. It is expressed in ms^{-1}.

When flow is *ideal*, the velocity across any cross sectional area is uniform; in other words, the mass flow rate across an area A under *ideal flow conditions* is given by:

$$\dot{m} = uA\rho \tag{2.1}$$

Some key theories developed for ideal flows will now be discussed.

2.2   Application of the Law of Conservation of Mass to Ideal Incompressible Flow

The law of conservation of mass was explained in this chapter. The same law will now be applied to an *ideal flow* (i.e. where the fluid offers no resistance to applied shear), with the additional assumption that the fluid is *incompressible*, i.e. the fluid density remains unchanged as it flows. It may be noted that the law of conservation of mass is general, and it is applicable to all flows. But the imposition of the above assumptions simplifies its mathematical formulation.

Consider the fluid entering an arbitrary conduit, shown in Fig. 2.3, at the end marked 2, and leaving at the end marked 1.

The rate at which the fluid mass enters at end 2 should be equal to the rate at which it leaves the conduit at end 1 for the validity of the law of conservation of mass. Since the mass flow rate is given by Eq. 2.1 for ideal flow, and the density ρ is the same at the two ends, the mass conservation principle yields,

$$\dot{m} = u_2 A_2 \rho = u_1 A_1 \rho; \text{ or } Q = u_1 A_1 = u_2 A_2 \tag{2.2}$$

Equation 2.2 is better known as the *continuity equation*, which, for an ideal incompressible flow, simply states that the product uA is constant; or $u \propto 1/A$. In other words, the superficial velocity is inversely proportional to the flow cross sectional area for a given flow rate. It is common practice to partially block the end of a garden hose with the finger in order to increase the flow velocity. The continuity equation provides the scientific explanation for this practice.

Fig. 2.3 Flow through an arbitrary conduit with the fluid entering the conduit at end 2, and leaving it at end 1

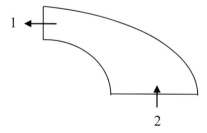

2.3 Bernoulli's Equation Applied to Ideal Incompressible Flow

Mechanical and thermal forms of energy are normally involved when fluids flow, and in many cases, the inter-conversion between these two forms can be significant. The application of energy conservation principle in its most general form will be too complicated at this stage, and this section will only consider the inter-conversion between different forms of mechanical energies involved in ideal incompressible flow. There are three forms of mechanical energy associated with such flows: *potential, kinetic and pressure energies*. Potential energy is the energy associated with position in the gravitational field and it is calculated as mgh where m is the mass and h is the position measured as height from a base line. The kinetic energy of a mass m is associated with the rate at which it moves and it is determined as: $\frac{1}{2}\,mu^2$. The pressure energy can simply be expressed as the product of pressure and volume. If a unit mass is now considered, the potential, kinetic and pressure energies are, respectively, given by: gh, $\frac{1}{2}\,u^2$ and P/ρ, where ρ is the density (note that the volume of unit mass is $1/\rho$).

Now consider ideal flow through a conduit such as the one shown in Fig. 2.3. If we equate the mechanical energies at ends 1 and 2, since it is conserved under the conditions imposed, we have:

$$gh_1 + \frac{1}{2}u_1^2 + \frac{P_1}{\rho} = gh_2 + \frac{1}{2}u_2^2 + \frac{P_2}{\rho} \tag{2.3}$$

It may be noted that h_1 and h_2 are the heights of ends 1 and 2 from an arbitrary baseline. Equation 2.3 is known as Bernoulli's equation, which can be re-written as follows by dividing both sides by g:

$$h_1 + \frac{u_1^2}{2g} + \frac{P_1}{\rho g} = h_2 + \frac{u_2^2}{2g} + \frac{P_2}{\rho g} \tag{2.4}$$

Each term of Eq. 2.4 has length units (i.e. m) and the three terms on each side of the equation are also known as *potential head*, *kinetic head* and *pressure head*, respectively. In terms of its physical significance, Eq. (2.4) states that the sum of the heads should be conserved, and one form of head can only change at the expense of others.

Fig. 2.4 Principle of an orifice meter

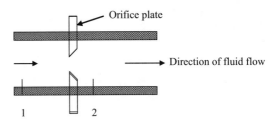

For example, if the flow occurs through a horizontal pipe (i.e. $h_1 = h_2$) and the velocity head increases, this increase can only occur at the expense of a drop in pressure head.

A simple instrument used to measure the volumetric flow rate in pipes, known as the *orifice meter*, operates on the above principle. A plate into which an orifice is drilled is inserted into the pipe, as shown in Fig. 2.4. The flow area is equal to the pipe cross sectional area at 1, and it contracts to a minimum value just after the fluid has gone past the orifice, say, at point 2; this position is also known as the *vena contracta*. The fluid velocity at end 2 is greater than that at end 1, and the two velocities are related by the continuity equation (Eq. 2.2). As a consequence of the increase in velocity, the pressure drops at end 2 well below that at end 1. The pressure difference between 1 and 2 is measured; and the flow rate is determined by applying Bernoulli's equation between the two points as follows:

$$\frac{u_1^2}{2} + \frac{P_1}{\rho} = \frac{u_2^2}{2} + \frac{P_2}{\rho} \tag{2.5}$$

By eliminating u_1 and u_2 between Eqs. 2.2 and 2.5, the flow rate is given by:

$$Q = A_2 \sqrt{\frac{2(P_1 - P_2)}{\rho\left[1 - \left(\frac{A_2}{A_1}\right)^2\right]}} \tag{2.6}$$

Since the area of the *vena contracta* (A_2) is unknown, it is customary to replace A_2 in Eq. (2.6) by A_0, the orifice area, and introduce an empirical correction factor, c_0, to restore the validity of the equation; i.e.,

$$Q = c_0 A_0 \sqrt{\frac{2\Delta P}{\rho\left[1 - \left(\frac{A_0}{A_1}\right)^2\right]}} \tag{2.7}$$

The correction factor c_0 is known as the *orifice discharge coefficient*, and its value depends on the design of the orifice and, under certain conditions, on the flow rate itself. In other words, it also includes irrecoverable energy losses associated with flow contraction and expansion. Manufacturers of orifice meters generally provide values of c_0, and these can be used to estimate flow rates when ΔP, the pressure drop across two points, is measured. Generally, $A_0 \ll A_1$, i.e. $A_0/A_1 \ll 1$; and Eq. (2.7) further simplifies to:

$$Q = c_0 A_0 \sqrt{\frac{2\Delta P}{\rho}} \tag{2.8}$$

Orifice meters are extensively used in industrial practice to measure flow rates through pipes, and it serves as an excellent illustration of the application of Bernoulli's equation.

Example 2.1 *1.25 litres per second of milk is to be pumped through a 25 mm diameter smooth stainless steel pipe, to an overhead tank 12 m higher. Estimate the power required and state clearly the assumptions made. The density of milk may be assumed to be 1050 kgm^{-3}.*

For the purpose of this problem, milk will be assumed to flow under ideal incompressible conditions, so that Bernoulli's equation can be applied in the form described by Eqs. (2.3) or (2.4). Further, we can assume that milk will flow at a uniform speed in the pipe and there are no kinetic energy effects. Thus, the potential head of 12 m has to be overcome by a pressure head developed by the pump. From Eq. 2.3:
$\frac{P_1-P_2}{\rho} = g(h_2 - h_1)$. Given h_2-h_1 = 12 m; $g = 10$ ms^{-2};and $\rho = 1050$ kg m^{-3}, the pressure difference developed by the pump works out to 126000 Pa. The power developed by the pump is equal to the product of pressure difference and volumetric pumping rate (1.25 lit/s or 1.25×10^{-3} m^{3}s^{-1}); i.e. $126000 \times 1.25 \times 10^{-3} = 157.5$ W. It may be noted that the ideal flow assumption excludes consideration of the friction between the fluid and the pipe wall. In any rigorous calculation, the frictional effects must be taken into account.

Example 2.2 *A liquid having density 1000 kg m^{-3} flows along a pipeline at a rate of 1.1 kg s^{-1}. The pipeline contains an orifice meter fitted with an orifice plate, 2.5 cm in diameter. A differential head is shown on a U tube manometer containing a manometric fluid having density 13600 kg m^{-3}. Determine the manometer reading if the orifice discharge coefficient is assumed to be 0.6.*

A manometer is used to measure pressure differential, and it normally consists of a U-tube, partially filled with a fluid that is immiscible with the flowing fluid. The two limbs of the U-tube are connected to pressure tapings located at the points across which the pressure differential is to be measured; see Fig. 2.5. The difference between the heights of the manometric fluid in the two limbs is measured (Δh_m); and the pressure drop is given by: $\Delta P = \Delta h_m(\rho_m - \rho_f)g$, where ρ_m and ρ_f are the

Fig. 2.5 An orifice meter set up with manometer connected. This figure is attached to Problem 2.2

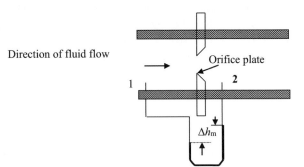

Fig. 2.6 A draining tank

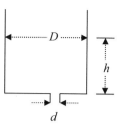

densities of the manometric and flowing fluids, respectively. A variety of manometers are available, and further information on their theory and practice can be found in (https://www.standardsuk.com/products/PD-ISO-TR-15377-2007/?gclid=CO6YwLLNutQCFRVmGwodo3IGJA).

In the above example, the mass flow rate of the liquid is given to be 1.1 kg s^{-1}; therefore its volumetric flow rate, Q = mass flow rate/density = 1.1/1000 or 1.1 × 10^{-3} m^3 s^{-1}. Further A_0 = π/4 (2.5 × 10^{-2})^2 = 4.9 × 10^{-4} m^2. Substituting the values of Q and A_0 in Eq. (2.8), ΔP = 6999.4 Pa. The manometer reading corresponding to this pressure drop can easily be calculated from the equation given in the above paragraph; i.e. $\Delta h_m = \Delta P/[(\rho_m - \rho_f)g]$ with ρ_m = 13600 kg m^{-3} and ρ_f = 1000 kg m^{-3}, to give Δh_m = 0.056 m or 5.6 cm.

Example 2.3 *How long would it take for a tank of diameter D and filled with a liquid to a height H to empty through a narrow circular hole of diameter d at its base?*

Consider the Fig. 2.6 shown below which illustrates the situation.

Let us apply Bernoulli's equation (Eq. 2.4), considering the surface and the base of the tank as the two positions. The liquid surface is at atmospheric pressure all the time as it drains; and the draining liquid also flows against atmospheric pressure at the base, which suggests that the pressure terms on both sides of Eq. (2.3) cancel out. The surface has potential head relative to the base of the tank, as well as a kinetic head as it drains. However, the kinetic head is negligible given the broad cross-sectional area of the tank (i.e. the speed at which its level drops is relatively small). On the other hand, the liquid flows out of the narrow hole at a significantly higher velocity v, giving it a dominant kinetic head at the base. Thus, the potential head at the surface is converted to kinetic head at the base. It must be noted that the surface potential head is itself falling all the time as the liquid level in the tank drops, as a consequence of which the kinetic head and the velocity at the base is progressively falling. In other words, the drainage rate is not uniform. Bernoulli's equation can now be used to quantify these arguments.

If the liquid height at any time, measured from the base of the tank, is h, then $h = h_2 - h_1$ in Eq. (2.4), where the surface is denoted by subscript 2 and the base by subscript 1. Therefore,

$$h = h_2 - h_1 = \frac{u_1^2 - u_2^2}{2g} \qquad (2.9)$$

Since $u_1 \gg u_2$, as discussed in the previous paragraph, the latter can be neglected in the above equation; and the liquid velocity at the base is simply given by:

$$u_1 = \sqrt{2gh} \qquad (2.10)$$

The above equation clearly shows that the drainage velocity at the base is proportional to the square root of the liquid height in the tank, which is consistent with the qualitative arguments proposed above.

We can now proceed to determine the time taken for the tank to empty. Applying continuity equation between the surface and base of the tank, Eq. (2.2) yields: $u_2 = u_1 A_2/A_1$. But u_2, the velocity of the surface, is also equal to the rate at which liquid level falls in the time, i.e. $\left(-\frac{dh}{dt}\right)$. Combining these results with Eq. (2.10), gives the following differential equation which describes the transient variation of liquid level in the tank:

$$-\frac{dh}{dt} = \frac{A_2}{A_1}\sqrt{2gh} = \frac{D^2}{d^2}\sqrt{2gh} \qquad (2.11)$$

Equation (2.11) can be solved with the initial condition that $h = H$ at $t = 0$, to give a relationship between the liquid level and time:

$$t = \frac{d^2}{D^2}\sqrt{\frac{2}{g}}\left(\sqrt{H} - \sqrt{h}\right) \qquad (2.12)$$

The time taken for the tank to empty can now be determined by allowing h in Eq. (2.12) to vanish; thus:

$$T = \frac{d^2}{D^2}\sqrt{\frac{2H}{g}} \qquad (2.13)$$

2.4 Flow of Real Fluids Through Pipes

The discussion above was restricted to ideal flows where fluids offered no resistance to applied shear. In practice, no flow is ideal in the true sense of the term, and *real flows* should consider frictional resistance, which is indicated by *viscosity*. As a consequence of viscosity, it is customary to assume that a *real* fluid does not *slip* at the pipe wall. In other words, the flow velocity at the wall is zero. This condition immediately suggests the existence of an axial velocity profile in pipes, which can have serious implications in food processing. For example, when fluids are thermally processed, they are heated to the processing temperature and held at that temperature for a given time. In continuous processes, the temperature is held by passing the fluid through a tube (known as the holding tube) at such a flow rate that the thermal effect is completed by the time the fluid leaves the pipe. The existence of a velocity profile creates a profile of residence times in the holding tube, which in turn creates a differential processing effect. For example, fluid elements traveling closer to the wall are moving slowly and tend to be over-processed in relation to the elements that are moving closer to the tube axis which are traveling significantly faster. An understanding of velocity profiles is therefore critical. Pipe flows can be broadly classified as being *laminar* or *turbulent*; and both these flow regimes are characterised by typical velocity profiles.

2.4.1   Laminar Flows

Laminar or *Streamline* flows generally prevail in small pipes, especially when fluid velocities are low, and they are characterized by the dominance of *viscous forces*. It follows that highly viscous fluids also tend to exhibit laminar flows. *Lamina* is a layer, and this flow can be characterised by imagining fluid layers stacked one above the other, flowing in an orderly manner. The axial velocity profile in the pipe is parabolic, with a maximum value occurring at the pipe axis which progressively decreases towards the wall where it becomes zero; see Fig. 2.7. The mean flow velocity, as explained in Sect. 2.1, is given by $u = Q/A = 4Q/\pi d^2$. The maximum velocity, which occurs at the axis, is twice the mean; i.e. $u_{max} = 2u$.

It was mentioned earlier in Sect. 2.4 that, in a real flow, the fluid offers resistance to applied shear stress. The effect of applied *shear stress* on a fluid is to produce a *shear rate*, and many common fluids, such as air and water, obey Newton's law which states that the shear stress and shear rate are proportional. The former is commonly denoted by the symbol τ and has units of Pa, while the latter is denoted by γ and has units of s^{-1}. Thus, according to Newton's law, $\tau \alpha \gamma$, or $\tau = \mu\gamma$, where μ is the constant of proportionality, known as *viscosity*. Thus Newton's law formally defines viscosity, and it has the units: Pas. It may be noted that water has an approximate viscosity of 10^{-3} Pas or 1 mPas under ambient conditions (around 18 °C). It is worth pointing out at this stage that not all fluids obey Newton's law.

Fig. 2.7 Velocity profile during laminar flow

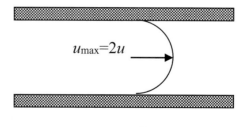

$u_{max}=2u$

$u=0$

Indeed a significant proportion of "biological" fluids do not strictly comply with this law, and "non Newtonian flows will be considered later in end 2. Until then we shall assume that the discussions are restricted to Newtonian flows. Indeed, the velocity profile and the characteristics of laminar flows discussed above are only valid when Newton's law applies.

2.4.2 Turbulent Flows

Inertial effects dominate turbulent flows. Although the bulk flow occurs in a given direction inside the pipe, the velocity at any position fluctuates with time. The instantaneous velocity at any position and in a given direction can be assumed to consist of two components – (i) a time averaged velocity in that direction and (ii) the fluctuating component. Thus, if $\overline{u_x}$ is the time averaged velocity in the x-direction and u'_x is the fluctuating component, the instantaneous velocity is given by:

$$u_x = \overline{u_x} + u'_x \qquad (2.14)$$

It may be noted that u'_x can assume any value – positive or negative; and it takes random values. It is a measure of the *intensity of turbulence*. If the fluctuating components of velocity in x-, y- and z-directions are equal, the turbulence is said to be *isotropic*. Prandtl postulated that in turbulent flow, the momentum is transferred by transient *eddies* which are lumps of fluid elements losing their identities as they mix into the fluid at another position (Coulson and Richardson 1999). These eddies essentially account for the velocity fluctuations at any point; and the mean distance they travel before losing their identity related to what is known as the *Prandtl mixing length*.

There are various theories describing the structure of turbulence. The theories which are extensively used in Computational Fluid Dynamics packages (commonly known as CFD) to simulate turbulent flows are based on the energy associated with turbulence and its rate of dissipation. In this section, we will not consider all the theories, but simply state that the effect of turbulence is to flatten the velocity profile observed in a pipe. Similar to laminar flow, the velocity is maximum at the pipe axis; and it progressively drops towards the wall where it becomes zero. However, unlike in the case of laminar flow, the maximum velocity is not twice the mean velocity, but approximately 1.2 times the mean velocity. This implies that the fluid velocity across the pipe axis is more uniform, as shown in Fig. 2.8.

Fig. 2.8 Velocity profile in turbulent flow, which is flatter than the profile for laminar flow shown in Fig. 2.7

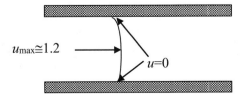

$u_{max} \cong 1.2$

$u = 0$

Due to the flatter velocity profile, the fluid residence time distribution has a narrower spread, which reduces differential heating effects. This, combined with the fact that heat and mass transfer rates are greater, has encouraged the extensive use of turbulent flows in industrial practice. It should however be noted that there are may food systems which are shear sensitive (e.g. yoghurt) and turbulent flows must be avoided to limit texture damage. Laminar flows, characterised by relatively lower shear, are therefore used in such cases.

2.5   Characterisation of Frictional Pressure Drop in Pipes and Fittings

2.5.1   Frictional Pressure Drop in Straight Pipes

Friction at a pipe wall results in shear stress (τ_w) which, in turn, generates a force that opposes fluid flow. If L is the pipe length and d is its diameter, as shown in Fig. 2.9, the wall shear force is equal to $\tau_w \pi dL$. Note that πdL is the area of the pipe wall that is in contact with the fluid. In writing this expression for wall shear force, we have also assumed that the shear stress is *uniformly* along the pipe length. In practice, the wall shear force is overcome by a pump which applies pressure at the upstream end, which progressively drops along the pipe length due to wall shear. If ΔP is the pressure drop across the length L, the force generated by the pump is the product of the pressure drop and the pipe cross sectional area; i.e. $\Delta P \pi d^2/4$. When the wall shear force and the force generated by the pump, balance one another, the fluid flows uniformly through the pipe.

Thus $\tau_w \pi dL = \Delta P \pi d^2/4$; or

$$\frac{\Delta P}{L} = \frac{4\tau_w}{d} \qquad (2.15)$$

The term $\Delta P/L$ is known as the *pressure gradient*; and it can be estimated if the wall shear stress, i.e. τ_w, is known. This estimation is very useful, because we will be able to specify pump characteristics (eg power rating) for any given flow condition. However, the wall shear stress is itself dependent on the flow conditions, and it

Fig. 2.9 Force balance in pipe flows

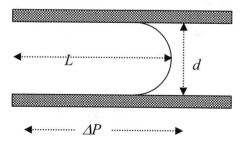

can be assumed to be proportional to the mean kinetic energy per unit volume of the flowing fluid; i.e. ½ ρu^2; note that u is the mean fluid velocity. Thus:

$$\tau_w = f\left(\frac{1}{2}\rho u^2\right) \tag{2.16}$$

where f is known as the *friction factor*, or more correctly, the Fanning friction factor. Substituting for τ_w from Eq. (2.16) into Eq. (2.15), the pressure gradient is given by:

$$\frac{\Delta P}{L} = \frac{2f\rho u^2}{d} \tag{2.17}$$

The value of the Fanning friction factor depends on whether the flow is laminar or turbulent. A simple method to find out the nature of flow in a pipeline is to estimate the *Reynolds number*, which is *dimensionless* and defined as follows:

$$Re = \frac{du\rho}{\mu}; \tag{2.18}$$

here μ is the fluid viscosity. This number can be shown to represent the ratio of inertial to viscous forces. Thus, high Reynolds number represents the domination of inertial force and hence turbulence, while low values represent the domination of viscous forces. If $Re < 2100$, the flow is laminar; and if $Re > 4000$, the flow is as turbulent. It is important to note that these limiting values are not absolutely rigid; and may be used as guidelines. For instance, if the Reynolds number is 4100, it cannot be assumed that the flow is fully turbulent. It can only be conjectured that the flow is not laminar and that turbulence is setting in. Indeed, fully developed turbulence, as is commonly understood, does not set in until Re values approach close to 10000. Regardless, the friction factor values have been correlated with Reynolds number and Fig. 2.10 shows how the two parameters are related. This chart should be consulted in order to estimate f for a given flow condition; the pressure gradient can then be determined from Eq. 2.17. Alternatively, empirical equations relating f with Re are also available for specified range of Re. For example, $f = 16/Re$ when the flow is laminar (i.e. $Re < 2100$); and $f = 0.08Re^{-0.25}$ when $5000 < Re < 100000$. The later equation – known as Blasius equation - is only valid for pipes having smooth walls. For reasons concerning food safety, manufacturers generally take extra care to ensure that the pipe walls are absolutely smooth and crevice free, so that the pipe does not harbour undesirable microorganisms. However, pipes can be rough, and appropriate corrections may have to be incorporated to determine friction factor. In general, the roughness is characterized by the so called roughness ratio (ε/d), which depends on the material of fabrication, as shown in Fig. 2.10.

Once the pressure gradient has been estimated using Eq. 2.17, the pressure drop across any pipe length can be determined and the *mechanical power* needed to enable the fluid to flow can be calculated as the product of the *pressure drop* and

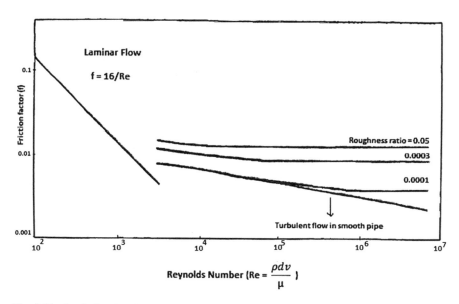

Fig. 2.10 Graph showing the Fanning friction factor as a function of Reynolds number for single phase Newtonian flows through smooth and rough pipes (Data taken from http://www.nzifst.org.nz/unitoperations/flfltheory5.htm). It is necessary to note that there are many ways of defining the friction factor, and there are many graphs relating Reynolds number with the corresponding friction factor. However, most friction factors are multiples of Fanning friction factor, and the nature of the graph is similar to the one shown above. Therefore, caution must be exercised while using such graphs to estimate pressure drop occurring in pipe flows. It may be noted that the data given in this figure can only be used when the pressure gradient is given by Eq. 2.17

volumetric flow rate. It should be noted that the *power rating* of the pump is the *electrical power* drawn by its motor; and this can easily be estimated if the efficiency of the pump in converting electrical power into mechanical power is known. This information is normally supplied by pump manufacturers.

Given that $f = 16/Re = 16\,\mu/du\rho$ for laminar flow, an expression for pressure gradient under laminar flow conditions can be derived by substituting this expression for f into Eq. 2.17; i.e.:

$$\frac{\Delta P}{L} = \frac{32\mu u}{d^2} \tag{2.19}$$

Equation 2.19 clearly demonstrates that $\Delta P/L$ is directly proportional to u and inversely proportional to d^2. A similar analysis can be undertaken for turbulent flow. Since f is weakly dependent on Re, to a very rough approximation, it can be assumed to be independent of Re. Therefore $\Delta P/L$ is directly proportional to u^2 and inversely to proportional to d when the flow is turbulent.

Example 2.4 *A horizontally mounted commercial steel pipe, 7 cm inside diameter, carries wastewater to a treatment plant, which is located 1.5 km away from the processing facility for hygienic reasons. The flow rate of the waste is 3 kg s^{-1}. Its density is 1050 kg m^{-3}, and viscosity is 0.001 Pas. Calculate the pressure drop due to skin friction and the power requirement for pumping if the pump used has an efficiency of 60%.*

Make it a habit to work in SI units. In this problem, $d = 7$ cm $= 7 \times 10^{-2}$ m; and the pipe cross sectional area, $A = \pi d^2/4 = 3.85 \times 10^{-3}$ m^2. The mass flow rate $=$ 3 kg s^{-1}; volumetric flow rate, Q $=$ mass flow rate / density $= 3/1050 = 2.86 \times 10^{-3}$ m^3 s^{-1}. The superficial velocity, $u = Q/A = 0.74$ m s^{-1}. The Reynolds number using Eq. (2.18), and the viscosity and density data given above $= 54600$, which indicates that the *flow is clearly turbulent*. The friction factor, using either the chart or the Blasius equation (i.e. $f = 0.08 \, Re^{-0.25}$), is equal to 5.23 x 10^{-3}. The pressure drop using Eq. (2.17) is given by:

$$\Delta P = \frac{2 \times \left(5.23 \times 10^{-3}\right) \times 1050 \times (0.74)^2 \times 1500}{0.07} = 1.29 \times 10^5 \, Pa$$

The mechanical power required for pumping $= \Delta P \times Q = (1.29 \times 10^5) \times (2.86 \times 10^{-3}) = 369$ W. Since the pump efficiency is 60%, the electrical power requirement $= 369/0.6 = 615$ W.

2.5.2 Pressure Drop in Pipe Fittings

All pipeline networks include valves and fittings, in addition to straight pipe sections. Valves essentially control the flow rate, while fittings perform a variety of functions depending on their design. For example: a *union* is used to join sections of straight pipe; *bends* are used to change flow directions; *expansion/contraction* joints are used to connect pipes of different diameters; and fittings in the shape of the letters *T* or *Y* are used to split flows into (or combine flows from) different pipe sections. Figure 2.11 illustrates commonly used valves and fittings.

Fluids suffer considerable pressure loss while passing through valves and fittings. Indeed, the combined pressure drop across valves and fittings invariably exceeds the pressure drop across straight sections by a substantial factor. The higher pressure loss is mainly due the nature of flow occurring within valves and fittings: the flow may undergo abrupt contractions and expansions, or abrupt changes in directions. Such changes result in a net loss of flow energy, which has to be borne by the pump. It may be noted that the fluid encounters two types of friction inside valves and fittings: one due to physical contact with the wall of the fittings, known as *skin friction* just as in the case of straight pipes, while the other due to the shape or form of the fittings, known more commonly as *form friction*. The latter is considerably greater and takes up most of the energy supplied to the fluid during flow.

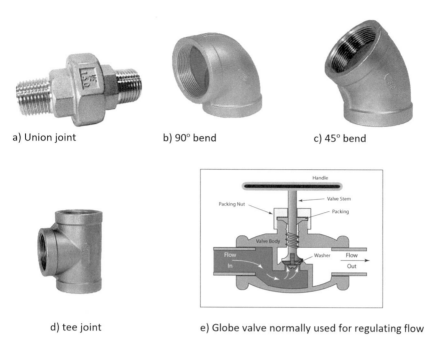

a) Union joint   b) 90° bend   c) 45° bend

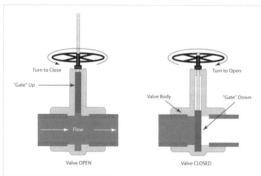

d) tee joint   e) Globe valve normally used for regulating flow

f) Gate valve with the gate shown in open and closed positions. This valve is normally used to switch the flow on or off.

Fig. 2.11 A selection of pipe fittings (http://www.haitima.com.tw/fitting.htm) and valves (https://www.quora.com/What-is-the-difference-between-a-gate-and-a-globe-valve). An exhaustive list of valves and fittings can be found by visiting the website of any reputed manufacturer. The number of velocity heads lost due to friction is approximately: union joint (negligible); 90° bend $\cong 0.6$–0.8; 45° bend $\cong 0.3$; tee joint $\cong 1.2$–1.8; fully open globe valve $\cong 1.2$–6; fully open gate valve $\cong 0.15$. The equivalent length values expressed in terms of pipe diameters are: union joint (negligible); 90° bend $\cong 30$–40; 45° bend $\cong 15$; tee joint $\cong 60$–90; fully open globe valve $\cong 60$–300; fully open gate valve $\cong 7$–10. (**a**) Union joint. (**b**) 90° bend. (**c**) 45° bend. (**d**) tee joint. (**e**) Globe valve normally used for regulating flow. (**f**) Gate valve with the gate shown in open and closed positions. This valve is normally used to switch the flow on or off

Two approaches are commonly used to characterise pressure drop across valves and fittings. One approach involves expressing the pressure drop in terms of the velocity head, i.e. $\Delta P = k\,(\rho u^2/2g)$ where k is the number of velocity heads lost due to the presence of the valve or fitting. Normally, k depends on the design and size of the valve or fitting, and in some cases, even on the flow Reynolds number. When the flow is highly turbulent, k depends on the former alone. It may be noted that the velocity u is the mean value based on the cross sectional area of the pipe.

A second approach is to define an *equivalent length of a straight pipe section*, which would offer the *same skin frictional pressure drop* as the valve or fitting for a given flow rate. The length is often expressed as *equivalent pipe diameters*. The equivalent for the fitting under consideration is then added on to the lengths of the straight pipe sections in the network to obtain an overall length, which is used instead of L in Eq. (2.17), to determine pressure drops. It may be noted that the equivalent length includes skin friction and shock losses, although the latter is dominant.

Example 2.5 *A pipe system having a nominal bore of 6 cm contains the following fittings with equivalent pipe diameters (EDP) as listed below:*

	EDP (per item)
4 × Gate valves	*10*
1 × globe valve	*200*
5 × 90° smooth elbows	*15*
2 × 90° square elbows	*30*

If the pipe run also contains 75 m of straight pipe, calculate the head loss due to friction and shock losses, when water flows through the system at 5 kg s^{-1}. Take the density of water to be 1000 kg m^{-3} and viscosity to be 1 mPas. For laminar flow, assume f = 16/Re, while for turbulent flow, assume f = 0.079Re$^{-0.25}$.

Given that the pipe diameter is 6 cm i.e. 6×10^{-2} m, the equivalent length of each gate valve is $10 \times 0.06 = 0.6$ m, and the net equivalent length due to all 4 gate valves is $4 \times 0.6 = 2.4$ m. Similarly, the net equivalent of all other fittings is as follows: globe valve: $200 \times 0.06 = 12$ m; 90° smooth elbows: $5 \times 15 \times 0.06 = 4.5$ m; and 90 square elbows: $2 \times 30 \times 0.06 = 3.6$ m. This gives a net equivalent length due to fittings of 22.5 m, which when combined with a straight run of 75 m gives a total equivalent length of 97.5 m.

The mass flow rate is 5 kgs^{-1} which is equivalent to 5×10^{-3} m^{3}s^{-1}, giving a superficial velocity $u = \frac{5 \times 10^{-3}}{\frac{\pi}{4} \times 0.06^2} = 1.77$ ms^{-1}. The Reynolds number can now be determined using Eq. (2.18) as 106200, which indicates that the flow is fully turbulent; and hence $f = 0.079(106200)^{-0.25} = 4.38 \times 10^{-3}$. Plugging relevant numerical values for all variables in Eq. (2.17),

$$\Delta P = \frac{2 \times 4.38 \times 10^{-3} \times 1000 \times 1.77^2 \times 97.5}{0.06} = 44596.8 \text{Pa}.$$ Since the head loss is equal to $\Delta P/\rho g$, its value is 4.55 m of water.

Table 2.1 Equivalent circular pipe diameters for common cross sectional shapes

Shape	Cross sectional area	Wetted Perimeter	Equivalent diameter
Square l	l^2	$4\,l$	l
Rectangle a b	Ab	$2(a + b)$	$\frac{2ab}{a+b}$
Open semicircle d	$\frac{\pi}{8}d^2$	$\frac{\pi}{2}d$	d
Closed semicircle d	$\frac{\pi}{8}d^2$	$\frac{\pi}{2}d + d$	$\frac{\pi d}{\pi+2}$

2.5.3 Pressure Drop in Pipes with Non-circular Cross Sections

The method described above assumes that the pipe cross section is circular. Although this is the case in most practical situations, pipes having other cross sectional shapes are also in use. For example, drains in processing areas are often open rectangular or semicircular channels; and flow in heat exchangers occur in tubes as well as through annular space between tubes. To estimate pressure drop in such cases, two points are noteworthy: 1. the flow velocity is based on the actual cross sectional area available for flow, and appropriate formulae should be used to determine this area; 2. the diameter in the definition of Reynolds number (i.e. Eq. 2.18) should be replaced by an *equivalent diameter or hydraulic diameter*, which is defined as (4 × *cross sectional area*) / *wetted perimeter*. Table 2.1 lists equivalent diameters for common cross sectional shapes.

Example 2.6 *A storage vessel is kept cooled by circulating a refrigerant ($\rho = 1100$ kg m^{-3}, $\mu = 14.0$ cP) through half-coils wound round the vessel as a jacket. The coils are of semi-circular cross section and made from 110 mm diameter rolled steel; the coils are equivalent to 120 m of straight pipe. If the refrigerant flows at a rate of 35 tonnes per hour, determine the power needed to pump it through the jacket.*

The volumetric flow rate of refrigerant, $Q = 35000/(1100 \times 3600) = 8.84 \times 10^{-3}$ m^{3}s^{-1}. The semicircular flow cross sectional area, $A = 0.5 \times \pi/4 \times d^2 = 0.5 \times \pi/4 \times (110 \times 10^{-3})^2 = 4.75 \times 10^{-3}$ m^2, which gives flow velocity, $u = Q/A = 1.86$ ms^{-1}. The equivalent diameter using the formula for a closed semi-circle given in Table 2.1 $= 0.5\pi d^2/(0.5\pi d + d) = 0.067$ m. The Reynolds number using Eq. (2.18) $= 0.067 \times 1.86 \times 1100/0.014 = 9792$. It may be noted that the viscosity has been given in centi Poise (cP) which is the same as mPas. The value of Re suggests that the flow is

Fig. 2.12 Optimum pipe diameter to be employed for a given flow rate

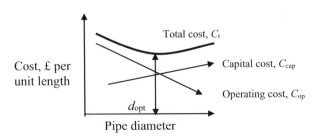

Cost, £ per unit length

Total cost, C_t

Capital cost, C_{cap}

Operating cost, C_{op}

d_{opt}

Pipe diameter

turbulent. The friction factor, f, from Fig. 2.10. is 7.9×10^{-3}. Substituting relevant values into Eq. 2.17, the pressure drop $\Delta P = 6.59 \times 10^4$ Pa. The power necessary to pump the refrigerant $= \Delta P \times Q = 553.89$ W.

2.5.4 The Concept of Economic Pipe Diameter

In Sect. 2.5.1, it was shown that, for a given flow rate, $\Delta P/L$ is inversely proportional to d^2 for laminar flow, while it is inversely proportional to d for turbulent flow. Regardless of the flow regime, it is clear that the pressure gradient will be lower if a larger pipe is used for a given flow rate. This suggests that the cost of pumping a given volume of fluid through a larger pipe is lower than that for a smaller pipe, indicating that larger pipes are more economical to run. Although this is the case, it must be borne in mind that larger pipes also cost more to buy; i.e. the capital investment will be high in the first place. It is therefore necessary to consider the economics carefully. Figure 2.12 shows on an arbitrary scale, how the *capital* and *operating* costs per unit pipe length vary as a function of pipe diameter. The *total cost* (i.e. the sum of capital and operating costs) goes through a minimum value at a pipe diameter known as the *optimum pipe diameter*. This diameter can be precisely determined by setting up models for capital and operating costs as a function of pipe diameter to deduce an expression for the total cost as: $C_t = C_{cap} + C_{op}$; the derivative $\frac{dC_t}{dd}$ is then set to zero to obtain d_{opt}.

Example 2.7 *The cost per meter of stainless steel pipe as a function of its diameter (also expressed in m) is given by $C_{cap} = 3156 \, d^{1.3} \, £ \, m^{-1}$. What is the optimum pipe diameter for carrying 5 kg s^{-1} of water under turbulent flow conditions if it is assumed that electrical power is available at 0.1£/kWh. The Blasius equation $f = 0.08Re^{-0.25}$ may be assumed to be valid for estimating the Fanning friction factor, f.*

The mass flow rate is 5 kgs^{-1} which is equivalent to a volumetric flow rate of 5×10^{-3} m^{3}s^{-1}, giving a superficial velocity $u = \frac{5 \times 10^{-3}}{\frac{\pi}{4}d^2} = 0.0063d^{-2}$ m s^{-1}. Substituting this expression for u in the expression for pressure gradient $\frac{\Delta P}{L} = \frac{2f\rho u^2}{d}$, with f being given by the Blasius equation, Re $= (du\rho/\mu)$, $\rho = 1000$ kgm^{-3} and $\mu = 0.001$ Pas, we get: $\frac{\Delta P}{L} = 0.04d^{-4.75}$. Since the mechanical power, P, required to pump the fluid trough a meter of pipe length is the product of the pressure gradient and the volumetric flowrate, $P = 2.004 \times 10^{-4}d^{-4.75}$ kWh. If we assume a pump motor efficiency of 60%, the electrical power supply will be $P/0.6 = 3.34 \times 10^{-4}d^{-4.75}$, which, at 0.1 £/kWh, gives an operating cost in £/m of $C_{op} = 3.34 \times 10^{-5}d^{-4.75}$. The total cost is therefore given by: $C_t = 3156d^{1.3} + 3.34 \times 10^{-5}d^{-4.75}$. The derivative $\frac{dC_t}{dd}$ must be set to zero in order to find the minimum total cost. When this is done and solved for d, we get $d_{opt} = 0.059$ m.

2.6 Power Law Flow Behaviour

Many flows encountered in food processing do not follow Newtonian behaviour which postulates that the shear stress (τ) is directly proportional to shear rate (γ). These include flow of fruits purees and concentrates, cocoa butter etc. A number of such flows (not all) can be explained on the basis of what is commonly known as power law: $\tau = k\gamma^n$, where k is known as the *consistency index* and n is known as the *power law coefficient* (or flow behaviour index). It is obvious that when n = 1, this equation reduces to indicate Newtonian flow. If the power law equation is re-written as: $\tau = k\gamma^{n-1}\gamma$, it is possible to define an *effective viscosity* by drawing a comparison with the Newtonian flow equation, i.e. $\tau = \mu\gamma$. Thus, $k\gamma^{n-1}$ takes the place of μ, and it is known as the effective viscosity; or $\mu_{eff} = k\gamma^{n-1}$. When n < 1, the exponent will be negative and μ_{eff} decreases with increasing shear. In other words, the fluid appears less viscous or "thinner" as the shear rate increases. Such flows are described as *shear thinning* or *pseudo plastic*. On the other hand, when n > 1, the exponent is positive and the flow is described as *shear thickening* or *dilatent*. The relationships between shear stress and shear rate for such flows are graphically illustrated in Fig. 2.13.

Figure 2.13 also demonstrates the relationship between shear stress and shear rate for the so called *Bingham plastic* flow, where the flow does not occur below a threshold shear stress (τ_o), known as the *yield stress*. Once the shear stress exceeds this value, the flow is essentially Newtonian. The only difference between the graphs for Newtonian and Bingham plastic flow, is that the former passes through the origin, while the latter has an intercept τ_o on the stress axis.

Apart from the ones shown in Fig. 2.13, it may be noted that food materials demonstrate many other types of non Newtonian behaviour, and a variety of

Fig. 2.13 Common types of power law flow behaviour

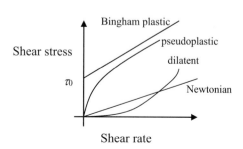

mathematical models are available to describe the relationship between shear stress and shear rate. Further, a significant proportion of food materials fall in the gray area between being a true solid and a true fluid. Such materials exhibit *viscous* (i.e fluid-like) *as well as elastic* properties (i.e. solid-like), when shear stress is applied, and are said to be *viscoelastic*. The study of the response of a material to applied shear is known as *rheology*, and the rheological characteristics of food materials play a key role in processing, as well as in determining texture, mouth-feel etc. Food rheology is a very well-developed subject, and a reader may refer to Rao (2007) for further information.

2.6.1 Pressure Drop During Power Law Flows

The method described in Sect. 2.5.1 to determine the pressure drop, i.e. Eq. 2.17, is essentially applicable to power law flows through straight pipes. The relationship between friction factor and Reynolds number is also similar. However, the definition of Reynolds number needs further consideration. Equation 2.18 defines the Reynolds number in terms of viscosity, μ. Such a definition cannot be directly applied to power-law flows, because the viscosity of the fluid is *not constant*; it depends on the shear level in the pipe, or the flow rate itself. In practice, a modified Reynolds number is defined for power-law flow represented by $\tau = k\gamma^n$:

$$\mathrm{Re}^* = \frac{d^n u^{2-n} \rho}{8^{n-1} k \left(\frac{3n+1}{4n}\right)^n} \tag{2.20}$$

With his definition of Reynolds number, the flow regime is deemed to be laminar if $\mathrm{Re}^* < 2100$, and approaching turbulent at greater values. The relationship between friction factor and modified Reynolds number is illustrated in Fig. 2.14, which shows that the relationship $f = 16/\mathrm{Re}^*$ continues to be valid for laminar power-law flows. For turbulent flows, on the other hand, f is also dependent on the value assumed by the power-law flow index (n). Power-law flows in food processing are more often laminar. This is particularly the case for shear sensitive materials which may undergo irreversible texture damage if subjected to the high shear levels encountered in turbulent flows. Regardless of the flow pattern, the pressure drop can be estimated from Eq. (2.17) provided an appropriate value of f is used. It may also be noted that the wall shear stress is still given by Eq. 2.15.

Example 2.8 *A power-law fluid ($\rho = 1000$ kg m^{-3}) gave the following results in a laboratory:*

| Shear stress (Nm^{-2}) | 15.5 | 31.0 | 67.0 | 104.0 |
| Shear rate (s^{-1}) | 10.0 | 30.0 | 100.0 | 200.0 |

The fluid is pumped at the rate of 15m^{3}h^{-1} through a pipe of diameter 7 cm and 80 m long. What is the pressure drop across the pipe?

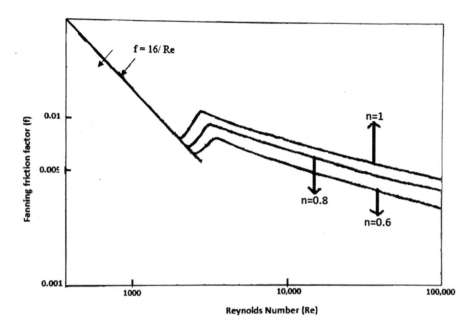

Fig. 2.14 Plot of Fanning friction factor against modified Reynolds number defined by Eq. 2.20 which enables the use of the same equation to estimate the Fanning friction factor for Newtonian and Power law flows in the laminar regime, i.e. f = 16/Re. (Data selected from Govier and Aziz (1987))

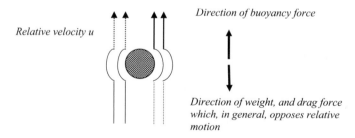

Fig. 2.15 Relative motion between a particle and flow, showing the forces acting on the particle. It may be noted that if the flow is downwards, the drag force will act upwards

The shear stress/shear rate data given above can be fitted to the power-law model and the best fit values of k and n are: 3.56 Pas^n and 0.64, respectively. The flow rate $= 15\text{m}^3\text{h}^{-1} = 4.17 \times 10^{-3}$ m^3 s^{-1}; and the pipe cross sectional area $= (\pi/4)d^2 = (\pi/4)(0.07)^2 = 3.85 \times 10^{-3}$ m^2, which gives a superficial velocity of 1.08 ms^{-1}. From Eq. (2.20), Re$^* = 110.51$, which indicates that the flow is laminar. Hence $f = 16/$Re$^* = 0.14$. The pressure drop using Eq. 2.17 is: 3.73×10^5 Pa.

2.7 Motion of a Particle in a Fluid (Fig. 2.15)

When a particle having volume V_p and density ρ_p moves *relative* to a fluid of density ρ_f, there are three forces acting on the particle:

1. the particle weight $= V_p\rho_p g$ which acts downwards;
2. the up-thrust or buoyancy force exerted by the liquid on the particle, which, according to Archimedes principle, is equal to the *weight of the fluid displaced* by it $= V_p\rho_f g$ (note that the particle will displace its own volume of liquid); and
3. a frictional force, F_d, known as the *drag force* which accounts for the abrupt changes in the flow direction around the particle, i.e. *from friction*, as well as *skin friction*, although, as mentioned in Sect. 2.5.2, the former dominates.

The drag force always acts in a direction that opposes motion. If the particle is rising through the fluid (e.g. a drop or a bubble), the drag force will act downwards, and if it is settling, the drag force will act upwards. In general, the drag force is proportional to two quantities: the kinetic energy per unit volume of the fluid, i.e. ½ ρu^2, and the particle area projected on to a plane perpendicular to flow, say A_p. Thus,

$$F_d = C_d A_p \frac{1}{2} \rho_f u^2 \qquad (2.21)$$

where C_d is a constant of proportionality known as the drag *coefficient*. The net force acting on the particle is therefore given by:

$$F = V_p\rho_p g - V_p\rho_f g \pm C_d A_p \frac{1}{2} \rho_f u^2 \qquad (2.22)$$

An assumption has to be made in respect of the sign of each term in Eq. 2.22: a force acting upwards can be taken to be positive, while a force acting downwards can be taken to be negative; and this convention can be followed to determine the sign of the last term. In general, the particle initially accelerates under the influence of F and gains speed. However, an increase in speed (u) also increases the drag force (i.e the last term), as a consequence of which, F falls and eventually vanishes. When this happens the particle assumes a uniform velocity, known as *terminal velocity*, u_t. Consider the case where the particle is *settling*, i.e. the drag force acts upwards, causing the last term to be negative. The terminal velocity can be deduced by setting $F = 0$ in Eq. 2.22, i.e.:

$$u_t = \sqrt{\frac{2V_p g\left(\rho_p - \rho_f\right)}{C_d A_p \rho_f}} \qquad (2.23)$$

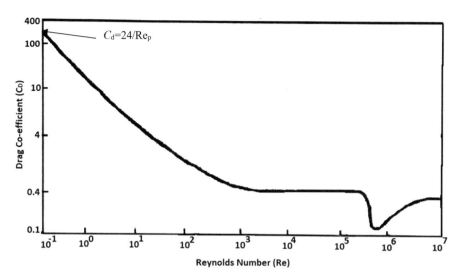

Fig. 2.16 Drag coefficient for a spherical particle as a function of Particle Reynolds number defined as $d_p u \rho_f / \mu_f$. (Data selected from Morrison (2013))

The above expression is also valid for the case where the particle *rises* through the fluid, except $(\rho_p - \rho_f)$ will be replaced by $(\rho_f - \rho_p)$ The drag coefficient, C_d, is related to the flow "disturbances" created by the particle, and it depends on the particle size and shape as well as the flow properties. A dimensionless number which takes into account all these variables is the *particle Reynolds number* defined as:

$$\mathrm{Re}_p = \frac{d_p u \rho_f}{\mu_f} \tag{2.24}$$

Here d_p is a characteristic length scale of the particle which depends on its size and shape. In the case of a sphere, d_p is its diameter. Further μ_f is the fluid viscosity. The drag coefficient has been experimentally determined as a function of Reynolds number, and the data for spheres is shown in Fig. 2.16.

Figure 2.16 states that $C_d = 24/\mathrm{Re}_p = 24\mu_f/(d_p u_t \rho_f)$ when $\mathrm{Re}_p \ll 1$. Under this condition, the flow around the particle is virtually streamline or laminar; and expressions for *drag force* and *terminal velocity* can be obtained by substituting this expression for C_d into Eqs. (2.21) and (2.23), respectively, after using the fact that $A_p = \frac{\pi}{4} d_p^2$ and $V_p = \frac{\pi}{6} d_p^3$. Thus:

$$F_d = \frac{6\pi d_p u_t \mu_f}{\rho_f} \tag{2.25}$$

$$u_t = \frac{g d_p^2 \left(\rho_p - \rho_f \right)}{18\mu_f} \tag{2.26}$$

Fig. 2.17 Flow through a
bed of spherical particles
having diameter d_p

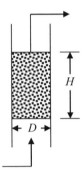

The drag force given by Eq. (2.25) is commonly known as Stokes law, and the
terminal velocity given by Eq. (2.26) is known as the Stokes terminal velocity. The
drag coefficient for non-spherical particles has also been studied extensively and a
comprehensive method to estimate it has been proposed by Holzer and
Sommerfield (2008).

2.8 Flow of a Fluid Through a Bed of Particles

Fluid flows through the voids or interstitial spaces formed between the particles in
the bed shown in Fig. 2.17. In terms of a model, the fluid can be assumed to flow
through hypothetical tubes formed by the particles, which are much smaller than the
column diameter. In fact the size of these tubes is closer to the particle size d_p. The
packed volume consists of two parts: (i) solid fraction and (ii) void fraction. The void
fraction of the bed, normally denoted by ε, is given by:

$$\varepsilon = \frac{V_{\text{total}} - V_{\text{solid}}}{V_{\text{total}}} = 1 - \frac{V_{\text{solid}}}{V_{\text{total}}} \tag{2.27}$$

The void fraction depends on the size and shape of the particles, and it can vary from
as low as 0.35 to 0.98; note it cannot be greater than 1.

In addition to the void fraction, a packed bed is also characterized by the specific
surface area, A_p, which is defined as the *particle surface area per unit volume of the
packed bed*. For a bed packed with spherical particles of diameter d_p, the total surface
area per unit bed volume is equal to $n\pi d_p^2$, where n is the number of particles per unit
volume; i.e.

$$A_p = n\pi d_p^2 \tag{2.28}$$

The total volume of all particles per unit bed volume is equal to $\frac{\pi}{6}nd_p^3$. If ε is the bed
voidage, the solid volume per unit bed volume must be equal to $(1 - \varepsilon)$. Thus:

Fig. 2.18 Variation of pressure gradient with velocity for flow through a packed bed

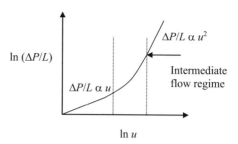

$$1 - \varepsilon = \frac{\pi}{6} n d_p^3 \quad \text{or} \quad n = \frac{6(1 - \varepsilon)}{\pi d_p^3} \tag{2.29}$$

Substituting the expression for n from Eq. (2.29) into Eq. (2.28), we have:

$$A = p \frac{6(1 - \varepsilon)}{d_p} \tag{2.30}$$

When a fluid moves through a packed bed, the tortuous flow results in form drag as well as skin frictional drag, but the former is significantly greater. As a consequence, the fluid suffers a pressure drop, depending on the flow conditions and fluid properties. Henry Darcy, a French engineer, studied flow through packed beds in the nineteenth century and reported that:

$$\frac{\Delta P}{L} \ \alpha \ \mu_f u$$

where μ_f is the fluid viscosity and u is its velocity. It was found that this relationship is only true for small particles and low flow rates; see Fig. 2.18. For a bed packed with spherical particles of diameter d_p, the pressure gradient has been experimentally found to be given by:

$$\frac{\Delta P}{L} = \frac{150(1 - \varepsilon)^2 \mu u}{\varepsilon^3 d_p^2} \tag{2.31}$$

where u is the fluid velocity based on the column cross sectional area. For high Reynolds number flow, the pressure gradient is proportional to u^2, and the pressure gradient is given by:

$$\frac{\Delta P}{L} = c \frac{(1 - \varepsilon)}{\varepsilon^3} \frac{\rho u^2}{d_p} \tag{2.32}$$

where c depends on the particle shape. For spheres, c takes the value 1.75. The pressure gradient for flow through spherical particles, in general, can be determined

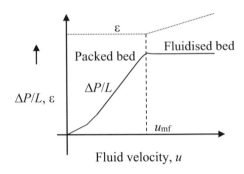

Fig. 2.19 Variation of pressure gradient and void fraction across packed beds against fluid velocity

by adding Eqs. (2.31) and (2.32) for a wide range of flow rates to yield the following equation which is more commonly known as the Ergun equation:

$$\frac{\Delta P}{L} = \frac{150(1-\varepsilon)^2 \mu u}{\varepsilon^3 d_p^2} + 1.75 \frac{(1-\varepsilon)}{\varepsilon^3} \frac{\rho u^2}{d_p} \tag{2.33}$$

It may be noted that the void fraction of the bed normally does not change with flow rate, and the particles remain well packed. However, there are limits. When the flow velocity reaches sufficiently high values, the particles start moving apart to make way for the fluid rushing past them. This causes the void fraction to increase, and the particles get suspended in the flow. As a consequence, over a narrow range of fluid velocity (or void fraction), the pressure gradient falls, as shown in Fig. 2.19, but soon picks up and remains relatively insensitive to pressure. The particles are now said to be *fluidized*, and the threshold velocity at which the particles just get suspended is known as the *minimum fluidization velocity* (u_{mf}). Above this velocity, the bed behaves like a fluid: the particles move about and mix just like a fluid, i.e. they flow, and indeed even obey laws that are normally obeyed by fluids, e.g. Bernoulli's theory.

Particles can be fluidized by gases as well as liquids. Normally, in liquid fluidized beds, the particles move further apart when the liquid velocity increases, and their motion becomes more vigorous. But the bed density at a given velocity is same in all sections of the bed. This behaviour is called *particulate fluidization* and it is characterized by a large but uniform expansion of the bed at high velocities. Air or gas fluidized beds, on the other hand, exhibit what is called *aggregative* or *bubbling fluidization*. At superficial velocities much greater than the minimum fluidisation velocity, most of the gas passes through the bed in the form of bubbles which are almost free of solids, and only a small fraction of the gas flows in the channels between the particles. It can indeed be postulated that the gas flow can be divided into two fractions: a flow corresponding to the minimum fluidisation velocity which fluidises the particles (i.e. creates the "fluid"), while the remaining gas – which is the dominant fraction – flows through the resultant "fluid" in the form of bubbles, just as they would rise in any liquid; see Fig. 2.20.

Fig. 2.20 Structure of: (**a**) particulate fluidized bed and (**b**) aggregative fluidized bed

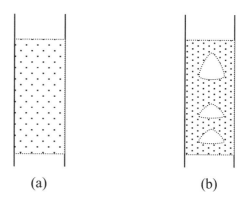

(a)   (b)

The chief advantage of fluidization is that the solid is vigorously agitated by the fluid passing through the bed, and the mixing of the solid ensures that there are practically no temperature or concentration gradients in the bed. The main disadvantages are: the requirement of larger volumes of fluid than what may be necessary, just to keep the particles fluidised, erosion of vessel internals, and attrition of particles. Fluidised beds find extensive application in food processing: for drying particles, mixing particulates, for spraying flavour or coatings on to particles, to carry out enzyme catalysed reactions where the enzyme is immobilised on a particle support, and in separation operations such as adsorption, chromatography etc.

Spouted bed is a variant of fluidized beds, mainly used when particles are too coarse and uniform in size for fluidization. Grains are normally spouted by air in order to dry. The fluid is forced up through a small orifice at the centre of a conical base, unlike in conventional fluidized beds where the fluid is evenly distributed at the base. Such a movement of fluid sets up a rapid upward movement of solids forming an axial spout, which fall into a relatively slow downward moving annulus. A sustained circulation of particles is set up, as shown in Fig. 2.21.

2.9 Concluding Remarks

This chapter essentially serves to introduce the basic concepts of fluid flow in pipes and through particles. This subject is very extensive and the topics covered here are limited to what one might call "basics". Flows encountered in food processing can be much more complicated than those described in this chapter, and the recommendations for further reading will inform a reader on topics such as compressible flows (i.e. flows where material density changes during flow) and multi-phase flows (gas-liquid, gas-solid and liquid-solid).

Fig. 2.21 Air spouted bed, commonly used for grain processing

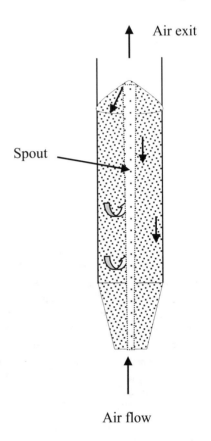

Air exit

Spout

Air flow

References

Coulson JM, Richardson JF (1999) Chem Eng 1:702
Govier GW, Aziz K (1987) The flow of complex mixtures in pipes, 2nd edn. Society of Petroleum Engineers. ISBN: 978-1-55563-139-0
Holzer A, Sommerfield M (2008) Powder Technol 184(3):361–365
Morrison FA (2013) An introduction to fluid mechanics. Cambridge University Press, New York, p 625

Further Reading

Chhabra RP, Shankar V (eds) (2017) Coulson and Richardson's chemical engineering volume 1A – Fluid flow fundamentals and applications, 7th edn
Rao MA (ed) (2007) Rheology of fluid and semisolid foods: principles and applications. Springer ISBN: 978-0-387-70929-1

Chapter 3
Elements of Heat Transfer

Keshavan Niranjan

Aims A vast majority of food processes involve either the addition of removal of heat. For instance, cooking and sterilisation involve heat addition, while chilling and freezing involve heat removal. In Chap. 1, thermal energy balances in such processes were briefly considered. However, there was no attempt to consider the factors which affected the rates at which heat could be added to (or removed from) foods. This chapter introduces the various modes of heat transfer, and discusses theories and models governing their rates.

3.1 Introduction

Heat transfer occurs when particles having *different temperatures* are either directly in contact, or brought into direct contact, or simply happen to be exposed to each other. *Conduction*, *convection* and *radiation* are recognised as the three modes by which heat can be transferred. Conduction refers to the transfer of heat from particle to particle within a material, without the bulk movement of the particles themselves. This type of heat transfer is common in solids; it may also dominate heat transfer in viscous liquids where bulk liquid movement may not be significant. For example, when meat is roasted in an oven, the heat penetrates from the surface inwards by conduction.

Convection, on the other hand, involves bulk material movement, which brings together particles having different temperatures. Bulk material movement (or convection) itself can be induced by temperature difference. For example, when a kettle containing water is heated by an electric heating element, the water

K. Niranjan
Department of Food and Nutritional Sciences, University of Reading, Reading, Berkshire, UK
e-mail: afsniran@reading.ac.uk

in contact with the element becomes hotter than the water further away from the element, as a consequence of which a density difference is developed causing the higher density water from farther regions to displace lower density water closer to the element thereby establishing circulation. When convection occurs as a consequence of density difference, it is known as *natural convection*, e.g. atmospheric air convection is natural! In food and other materials that are in the fluid state, it is also possible to *force* movement, by agitating or pumping the materials. This is common industrial practice, and such movements are known as *forced convection*. When materials having different temperatures get mixed, either "naturally" or "forcibly", heat transfer occurs by convection, and accordingly, the process is known as *natural convective* or *forced convective* heat transfer.

Oven baking is another example of a process using convection. Ovens, typically, contain two heating elements, one at the top and the other at the bottom. During baking, the bottom element heats up the air inside the oven, which rises, circulates and distributes heat throughout the oven. Such natural convection currents are easily blocked by large pans, creating non-uniform temperatures within the oven. The so-called convection ovens improve temperature distribution by using a fan located inside the oven, which forces convection currents. The increased airflow enhances heat transfer rates and reduces cooking time.

Unlike the above two modes of heat transfer, both of which involve direct contact between the particles exchanging heat, radiative heat transfer occurs when particles having different temperatures are simply exposed to one another, regardless of whether there is a medium separating them or not. For example, heat from the sun reaches the planets by radiation. Even in food processes, radiation can play a significant role: for example, the heat reaching a piece of meat in a hot oven can largely be transmitted by radiation from the heating element.

Two points may be noted from the above discussion on modes of heat transfer. First of all, the existence of a temperature difference is a pre-requisite for *net* heat transfer to occur. Secondly, in any practical process, no single mode of transfer operates in isolation. For example, a piece of meat gets cooked in an oven because heat from the heating element reaches its surface by a combination of radiation and air convection, and penetrates inside it by conduction. Thus all three modes are simultaneously involved. However, in the analysis of the *rate* at which the temperature at the centre of the meat rises, only one, or two of the three modes may be critical. If the rate of temperature rise is regarded as the net result of two or more processes occurring *in series*, e.g. in the case of oven heating of meat, the heat transfer from the heating element to the surface by radiation and air convection, followed by internal transmission by conduction, the conductive step is far slower in relation to the former, and tends to control the overall rate. *It may be noted that the slowest of a series of steps involved in any rate process, controls the overall rate.*

3.2 Conductive Heat Transfer

The mathematical description of the rate of heat transfer is based on Fourier's law, which can most easily be expressed by considering a long thin rod: the rate of heat transfer per unit area past any plane perpendicular to the length of the rod, also known as the heat flux (Wm^{-2}), is proportional to the temperature gradient at any axial positon x, i.e. dT/dx (Km^{-1}). Thus,

$$q \ \alpha \ \frac{dT}{dx} \ \text{or} \ q \ = -\lambda \ \frac{dT}{dx} \tag{3.1}$$

where λ is the constant of proportionality known as the *thermal conductivity*. The negative sign has been introduced to make both sides positive on the whole, since the temperature gradient is inherently negative. In other words, if heat is flowing in the x-direction, or in the direction of increasing x value, the temperature has to fall in this direction since heat flows only from a higher to a lower temperature. Further, since the thermal conductivity can be looked upon as the heat flux per unit temperature gradient, it has the following units: $\mathrm{Wm}^{-1}\,\mathrm{K}^{-1}$. Table 3.1 lists thermal conductivity values of common materials used in food processing. It may be noted that the thermal conductivity, like other thermal properties, can vary with temperature, and it often increases with temperature. If the temperature gradient across a rod of length L is uniform (see Fig. 3.1), the heat flux, given by Eq. 3.1, can also be written as:

Table 3.1 Density (ρ), thermal conductivity (λ), specific heat (C_p) and thermal diffusivity $(\alpha = \lambda/[\rho C_p])$ of some common materials encountered in food processing

	$\rho(\mathrm{kg\,m}^{-3})$	$\lambda(\mathrm{Wm}^{-1}\,\mathrm{K}^{-1})$	$C_p(\mathrm{J\,kg}^{-1}\,\mathrm{K}^{-1})$	$\alpha \times 10^6 (\mathrm{m}^2\,\mathrm{s}^{-1})$
Aluminum	2780	170	880	70
Copper	8900	400	385	117
Stainless steel	8000	15	480	3.7
Glass	2600	4	800	1.9
Water	1000	0.5	4200	0.12
ice	920	2.2	2050	1.1

The properties of liquid foods like milk and fruit juices would be expected to be close to those of water. It may be noted that the purpose of this table is to indicate some typical values of the properties one can expect to find. The values are therefore not accurate, and better values – which are temperature dependent – can be obtained from published literature

Fig. 3.1 Uniform heat flux through heat conducting rod AB, showing linear variation of temperature with length

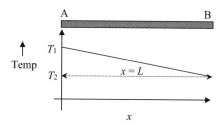

$$q = -\lambda \frac{T_2 - T_1}{L} \qquad (3.2)$$

where T_1 and T_2 are the temperatures at the two ends of the rod ($T_1 > T_2$). It is worth noting that the heat flux in the rod can only be uniform if the heat entering the rod at A is quantitatively removed from it at the same rate, at the end B. As a consequence, the heat flux (q) and the temperature (T) at any position x are independent of time, and the heat transfer is said to occur under steady state conditions.

An alternative approach to describe heat flux is to assume that it is directly proportional to the *temperature difference*, which is the driving force *for heat transfer*. In other words,

$$q\alpha(T_1 - T_2) \text{ or } q = h(T_1 - T_2) = \frac{(T_1 - T_2)}{1/h} \qquad (3.3)$$

where h, the constant of proportionality, is known as the *heat transfer coefficient*, and has units ($Wm^{-2} K^{-1}$). If we set T_1-$T_2 = 1$ K, then $q = h$, i.e. the heat transfer coefficient can be thought of as being equal to the flux per unit temperature driving force. Moreover, if any flux - be it relating to the transfer of heat or electricity or mass – can be looked upon as a driving force for transfer, divided by the resistance to transfer, $1/h$ represents the resistance to heat transfer. Comparing Eqs. (3.2 and 3.3), we have $h = \lambda/L$, or:

$$\frac{hL}{\lambda} = 1 \qquad (3.4)$$

The left hand side of Eq. 3.4 represents a dimensionless number and it is known as the *Nusselt number, Nu*. For the situation considered in Fig. 3.1, i.e. for unidirectional heat conduction through a rod, $Nu = 1$.

If, instead of a rod, we considered a slab of material with two parallel faces separated by thickness L (Fig. 3.2), the Nusselt number will still be given by the same value, i.e. $Nu = hL/\lambda = 1$; and the flux will be given by Eq. 3.2. Furthermore, if $1/h$ is a measure of the resistance offered by the slab to heat transfer, so will L/λ. If the slab is a composite of two or more materials placed one on top of the other, the overall resistance will be the sum of the individual resistances (since these resistances are in series), and the heat flux will be given by:

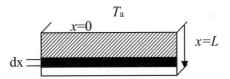

Fig. 3.2 Heat conduction through an infinite slab of thickness $x = L$ exposed to a temperature T_a at $x = 0$; dx represents the thickness of an element of the slab

$$q = \frac{(T_1 - T_2)}{L_1/\lambda_1 + L_2/\lambda_2 + L_3/\lambda_3 + \ldots}$$ (3.5)

where $L_1, L_2, L_3 \ldots$ etc. are the thickness of the component slabs and $\lambda_1, \lambda_2, \lambda_3 \ldots$ etc. are their respective thermal conductivities.

This argument can also be extended to radial heat conduction through a sphere of radius R, and the Nusselts number in this case is given by $hR/\lambda = 1$. It is however more common to express the Nusselt number for a sphere in terms of diameter D, which gives:

$$Nu = \frac{hD}{\lambda} = 2$$ (3.6)

The theory of steady state conduction can be used to determine freezing time of foods. For instance, if T_a in Fig. 3.2 is the ambient temperature in a freezer, q is the flux, and a thickness x has already frozen at temperature T_f, the rate at which mass is frozen is given by:

$$\frac{dm}{dt} = \frac{qA}{h_f}$$ (3.7)

where h_f is the latent heat of fusion and A is the cross sectional area of the slab.. This mass will also be in the shape of a slab, but it will have a thickness of dx; i.e. $m = Adx\rho$. In other words,

$$\frac{dm}{dt} = A\rho \frac{dx}{dt}$$ (3.8)

Combining Eqs. (3.7) and (3.8), we have

$$\frac{q}{h_f} = \rho \frac{dx}{dt}$$ (3.9)

Noting that $q = h(T_f - T_a)$, where $h = \lambda/x$ according to Eq. (3.4), we have:

$$\frac{dx}{dt} = \frac{\lambda(T_f - T_a)}{\rho h_f x}$$ (3.10)

The above differential equation can be solved with the initial condition, $x = 0$ at $t = 0$, to give:

$$\frac{x^2}{2} = \frac{\lambda(T_f - T_a)}{\rho h_f} t$$ (3.11)

The freezing time can easily be deduced by putting $x = L$ in the above equation; i.e.:

$$t_f = \frac{L^2 \rho h_f \left(T_f - T_a\right)}{2\lambda} \tag{3.12}$$

The above model assumes that the slab is at its melting or freezing point throughout the melting process, and that the flux is uniform.

To a certain approximation, the thawing of a frozen slab can also be treated as a steady-state conduction problem where heat entering the slab, which is at its freezing point, is quantitatively used to melt it. With reference to the slab shown in Fig. 3.2, T_a will be higher than the temperature of the slab which is at its melting point T_f, and heat will be conducted through the molten layer (i.e. rectangular region of thickness x, filled with light upward diagonal), to the dark shaded rectangle, thickness dx, which will melt due to the thermal flu3. Based on this model, Eq. (3.12) can also be used to estimate the thawing time, except that $(T_f - T_a)$ will be replaced by $(T_a - T_f)$. There is however a key point to be noted: in the case of freezing, the heat is conducted across a frozen layer and therefore λ *represents the thermal conductivity of the frozen material*; while in the case of thawing, the heat is transferred through a molten layer, and λ *represents the thermal conductivity of the molten liquid*. Since liquids generally have a significantly lower thermal conductivity than solids, *it takes much longer to thaw a material than to freeze it*.

In all the situations considered above (i.e. Figs. 3.1 and 3.2), the thermal flux has been assumed to be uniform; the temperatures at various points remain unchanged with time; and conductive heat transfer occurs under steady state conditions. Let us now consider the case shown in Fig. 3.3, where a cylindrical rod of *unit* cross sectional area is supplied with heat at end A, as a result of which the rod heats up along its length, and temperatures at various axial positions change with time. The thermal flux will therefore vary along the axis, decreasing progressively as one moves away from the heated end (i.e. in the direction of increasing x). This type of heat transfer is described as *unsteady state heat conduction*.

If $(-dq)$ is the change in thermal flux (i.e. rate of heat transfer in this case since the rod is of unit cross sectional area) over an elemental length of the rod, dx, causing a temperature rise dT in a time interval dt, a heat balance around this section can be drawn up as follows: the loss in heat flux is equal to the rate of heat absorbed by the elemental section of the rod, i.e.

$$-dq = mC_p \frac{\partial T}{\partial t} = \rho dx C_p \frac{\partial T}{\partial t} \tag{3.13}$$

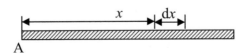

Fig. 3.3 Unsteady state conduction through a cylindrical rod of unit cross sectional area heated at end A

It may be noted that the mass of the elemental section of the rod, m, in the above equation has been replaced on the right hand side, by the product of density, ρ, and the volume of the section, $A dx$, where A, the rod cross sectional area, which will be assumed to be equal to 1 m^2 in further discussions. It is also essential to note that partial derivative signs have been used to represent the rate of change of temperature, since temperature in this situation is a function of two parameters, t and x. Since q is given by the Fourier's law, i.e. Eq. 3.1, we have,

$$-d\left(-\lambda\frac{\partial T}{\partial x}\right) = \rho dx C_p \frac{\partial T}{\partial t} \qquad (3.14)$$

which can be rearranged to give,

$$\frac{\partial T}{\partial t} = \frac{1}{\rho C_p}\frac{\partial}{\partial x}\left(\lambda\frac{\partial T}{\partial x}\right) \qquad (3.15)$$

If thermal conductivity, λ, is assumed to be independent of temperature, Eq. (3.15) can be simplified to yield:

$$\frac{\partial T}{\partial t} = \frac{\lambda}{\rho C_p}\left(\frac{\partial^2 T}{\partial x^2}\right) = \alpha\left(\frac{\partial^2 T}{\partial x^2}\right) \qquad (3.16)$$

In the above equation, $\alpha = \lambda/\rho C_p$ is known as the *thermal diffusivity*, and it has units of m^{2}s^{-1}. Thermal diffusivity measures the effectiveness by which a material conducts thermal energy with respect to its ability to store thermal energy. A material with high α is characterized by a quick response to the changes in surrounding temperatures. A material with low α takes longer to reach a steady state condition, but is excellent at retaining heat once heated. Typical values of α for common materials used in food processing are listed in Table 3.1.

It may be noted that Eq. (3.16), although derived for the case of a rod, is a general equation for unidirectional conductive heat transfer. It may be applied to different geometries, such as a slab (Fig. 3.2) or a cylinder or a sphere, *provided temperature varies temporally and in one spatial direction.*

Equation (3.16) is a partial differential equation which must be solved if the temperature at any position on the rod at any time is to be known. It is also evident that the differential equation is *first order* with respect to time and *second order* with respect to x. Therefore three boundary conditions have to be specified in order to obtain the solution: two conditions relating T with x, and one relating T with t. Graphical and other forms of solutions for Eq. (3.16) as well as for differential equations valid for other shapes, are described in a treatise on this subject authored by Carslaw and Jaeger (1959).

One possible condition for the rod shown in Fig. 3.1 is a constant temperature of T_A at end A. In this case, the temperature inside the wall at time t, at distance x from the surface, is given by:

Table 3.2 Values of Gaussian error functions: $erf(x) = \frac{2}{\sqrt{\pi}} \int_0^x e^{-t^2} dt$; and its complementary function $erfc(x) = 1 - erf(x)$

x	$erf(x)$	$erfc(x)$	x	$erf(x)$	$erfc(x)$
0.0	0.000	1.000	1.3	0.934	0.066
0.1	0.112	0.888	1.4	0.952	0.048
0.2	0.223	0.777	1.5	0.966	0.034
0.3	0.329	0.671	1.6	0.976	0.024
0.4	0.428	0.572	1.7	0.984	0.016
0.5	0.520	0.480	1.8	0.989	0.011
0.6	0.604	0.396	1.9	0.993	0.007
0.7	0.678	0.322	2.0	0.995	0.005
0.8	0.742	0.258	2.1	0.997	0.003
0.9	0.797	0.203	2.2	0.998	0.002
1.0	0.843	0.157	2.3	0.999	0.001
1.1	0.880	0.120	2.4	0.999	0.001
1.2	0.910	0.190	2.5	1.000	0.000

Although the complementary error function does not appear in Eq. 3.17, it does appear in solutions of Eq. 3.16 with other boundary conditions

$$\frac{\theta}{\theta_i} = \frac{T(x,t) - T_A}{T_i - T_A} = erf\left(\frac{x}{2\sqrt{\alpha t}}\right) \tag{3.17}$$

where α is the *thermal diffusivity* of the material of the rod. The function *erf* is called the *Gaussian Error Function*, and values for *erf*(x) are tabulated in Table 3.2.

In many practical situations, the end A is subjected to convection, which alters the boundary condition, and therefore the solution. Such situations will be discussed later in Sect. 3.5.

Problem 3.1 A thick slab of flat bread dough, initially at 20 °C throughout, is placed on a hot pan and it immediately attains the constant pan temperature of 120 °C. What is the temperature 2.5 mm from the pan, 20s after placing the dough sheet? The thermal diffusivity of the dough may be assumed to be the same as that of water, stated to be 1.2×10^{-7} m^2 s^{-1} in Table 3.1.

The answer to this question involves the direct application of Eq. 3.17, i.e.: $\frac{T(x,t)-T_A}{T_i-T_A} = erf\left(\frac{x}{2\sqrt{\alpha t}}\right)$, where $x = 2.5 \times 10^{-3}$ m, $t = 20$s, $T_A = 120$ °C, $T_i = 20$ °C and $\alpha = 1.2 \times 10^{-7}$ m^2 s^{-1}. The term within brackets of the error function is evaluated to be 0.807 and the value of the error function may be approximated to 0.75. Therefore, the value of T (2.5 mm, 20 °C) = 45 °C.

It is necessary to note in the above problem that we have assumed the dough to be a homogeneous solid through which heat is conducted. This is far from true because the dough contains water, which can potentially evaporate at the surface of the flat bread where it is in contact with the pan, and the temperature pattern in the slab of dough may well be determined by the movement of steam and the heat it loses as it comes into contact with the colder layers of the dough, which is a significantly more complex situation. Thus, it is an oversimplification to assume that a single mechanism of heat transfer operates in food materials.

3.3 Convective Heat Transfer

3.3.1 Mechanism of Convective Heat Transfer

Heat transfer in fluids is dominated by convective mechanisms. Typical situations encountered in practice include heat transfer to or from liquids processed in mixing vessels, or flowing in pipes. It is the movement of the fluid that causes various elements, having different temperatures, to come into contact and exchange heat.

The mechanism of heat transfer by convection can be understood from the model shown in Fig. 3.4. The fluid contained in the jacket is at a temperature T_w, loses heat to the fluid inside the vessel (referred to as the process fluid), which is at a temperature T_m, through the vessel wall. Let us, first of all, consider the temperature profile (Fig. 3.4b). Since heat flows in the direction of decreasing temperature, the fluid inside the jacket loses heat to the face of the vessel wall with which it is in contact. This face should therefore be at a lower temperature, say T_{wo}. The heat then flows across the wall to the face which is in contact with the process fluid. Since the flow of this heat also requires a temperature driving force, the temperature of the inner face of the wall has to be lower than T_{wo}, say T_{wi}. Now, the heat is transferred from the inner face of the wall to the process fluid, which has to posses an even lower temperature, i.e. T_m. Thus, the temperature profile follows: $T_w > T_{wo} > T_{wi} > T_m$. Let us now consider the mechanism by which heat is transferred from the inner face of the vessel wall to the process fluid. The fluid in contact with the inner face of the

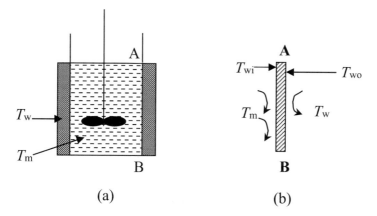

(a)   (b)

Fig. 3.4 (**a**) Convective heat transfer from a heating fluid at temperature T_w inside a jacket, to a process fluid inside a vessel at a temperature T_m. (**b**) Region around wall AB is shown in greater detail. T_{wo} is the temperature of the outer face of the wall AB and T_{wi} is the temperature of the inner face. Given the direction of heat transfer, $T_w > T_{wo} > T_{wi} > T_m$

vessel wall absorbs heat from it and acquires its temperature. Given that the impeller is constantly mixing the process fluid, i.e. the rotating impeller creates convection currents in it, the fluid elements in contact with the inner face of the vessel wall are swept into the bulk where they give up heat to other elements with which they come into contact. The same convective action also replenishes the inner face with fresh fluid from the bulk, thereby sustaining the process.

Three steps, each with a different temperature driving force, can now be identified which combine to effect the overall heat transfer: 1. *convective* heat transfer from the inner face of the vessel wall to the process fluid under the driving force $(T_{wi} - T_m)$, 2. *conductive* heat transfer across the vessel wall under the driving force $(T_{wo} - T_{wi})$, and 3. yet another convective heat transfer between the fluid in the jacket and the outer face of the vessel wall under the driving force $(T_w - T_{wo})$. Following the discussion in Sect. 3.2 leading to Eq. 3.3, the heat flux under convection can also be assumed to be proportional to the temperature driving force, which yields the following equations for the rate of heat transfer for each of the above steps:

$$Q_1 = h_1 A_1 (T_{wi} - T_m) \tag{3.18}$$

$$Q_2 = h_2 A_2 (T_{wo} - T_{wi}) = \frac{\lambda}{x} A_2 (T_{wo} - T_{wi}) \tag{3.19}$$

$$Q_3 = h_3 A_3 (T_w - T_{wo}) \tag{3.20}$$

In the above equations, A_1, A_2, and A_3 are the areas for heat transfer, and h_1, h_2 and h_3 are the respective heat transfer coefficients. It may also be noted that Eq. (3.19) represents conduction through the vessel wall; and h_2 has been equated with λ/x, where x is the wall thickness, following Eq. 3.4. The values of the convective heat transfer coefficients h_1 and h_3 depend on fluid convective movements and this will be discussed later. For the moment, let us consider how each stage affects the overall rate of heat transfer. Given that the three stages are in series, the same rate of heat transfer must occur through each stage at steady state; i.e. $Q_1 = Q_2 = Q_3 = Q$ where Q is the overall rate of heat transfer. If the overall rate of heat transfer is expressed in terms of an overall heat transfer coefficient U as follows:

$$Q = UA_1 (T_w - T_m) \tag{3.21}$$

it can easily be shown that:

$$\frac{1}{U} = \frac{1}{h_1} + \frac{1}{h_2} \left(\frac{A_1}{A_2} \right) + \frac{1}{h_3} \left(\frac{A_1}{A_3} \right) \tag{3.22}$$

The overall heat transfer coefficient defined in Eq. (3.21) can also be based on either A_2 or A_3, and accordingly, equations similar to Eq. (3.22) can be deduced. In many practical situations, the values of A_1, A_2 and A_3 are not very different, i.e. the ratio of the areas is approximately 1, and Eq. (3.22) simplifies to:

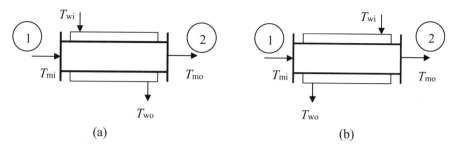

Fig. 3.5 Temperature differences for flow through a pipe fitted with jacket. (**a**) co-current config-uration: $\Delta T_1 = (T_{wi} - T_{mi})$; $\Delta T_2 = (T_{wo} - T_{mo})$. (**b**) counter-current configuration: $\Delta T_1 = (T_{wo} - T_{mi})$; $\Delta T_2 = (T_{wi} - T_{mo})$

$$\frac{1}{U} = \frac{1}{h_1} + \frac{1}{h_2} + \frac{1}{h_3} = \frac{1}{h_1} + \frac{x}{\lambda} + \frac{1}{h_3} \qquad (3.23)$$

If we recall that the reciprocal of heat transfer coefficient is a measure of the resistance to heat transfer (Sect. 3.2), Eq. (3.23) simply states that the overall resistance is equal to the sum of individual resistances, which is well known to be the case when resistances in any process act in series.

In the situation described above, the temperatures of the heating and process fluids, i.e. T_w and T_m are constant over the entire heat transfer area. This is because both the fluids are assumed to be well mixed. If heat transfer occurs between a heating fluid and a process fluid in a pipe fitted with a jacket, there are two possible flow configurations as shown in Fig. 3.5a, b: *co-current* configuration where the process and heating fluids flow in the same direction, and *counter-current* configu-ration where the process and heating fluids flow in opposite directions. The temper-ature driving force for heat transfer changes along the length of the pipe in both the cases. In such situations, the average driving force is taken as the logarithmic mean temperature difference, defined as:

$$\Delta T_{ln} = \frac{\Delta T_1 - \Delta T_2}{\ln \left(\frac{\Delta T_1}{\Delta T_2}\right)} \qquad (3.24)$$

where ΔT_1 and ΔT_2 are the driving forces at the two ends of the pipe; and the rate of heat transfer given by Eq. (3.24) changes to:

$$Q = UA\Delta T_{ln} \qquad (3.25)$$

It may be noted that Eq. (3.24) is valid regardless of whether the two fluids flow in the co- or counter-current mode, although the numerical values of ΔT_1 and ΔT_2, and hence ΔT_{ln}, will be different for the two cases. In general, the log mean temperature difference takes values between the arithmetic and geometric means of the two temperature differences.

Normally, counter-current configuration is preferred when heat exchange between the process and heating fluids involves sensible heat changes in both fluids. When phase change occurs in either fluids, e.g. the heating medium is condensing steam, the log mean temperature difference is not significantly different between the two configurations, and either modes of operation can be used. For reasons relating to energy efficiency, it is desirable to maintain a local temperature difference between the two fluids of at least 10 °C. In practice, this may not be possible because high temperature difference tends to promote chemical reactions in foods which deposit a fouling layer on the heat exchange surface. This can adversely affect heat transfer rates, because the fouling layer – which is a solid deposit of low thermal conductivity – offers an additional resistance to heat transfer, acting in series with the other resistances described above (i.e. in the paragraph leading to Eqs. (3.22) and (3.23)). Thus, when fouling has occurred in a heat exchanger, the overall heat transfer coefficient can be determined by modifying Eq. (3.23) as follows:

$$\frac{1}{U} = \frac{1}{h_1} + \frac{1}{h_2} + \frac{1}{h_3} + \frac{1}{h_{\text{fouling}}} = \frac{1}{h_1} + \frac{x}{\lambda} + \frac{1}{h_3} + \frac{x_{\text{fouling}}}{\lambda_{\text{fouling}}} \qquad (3.26)$$

where x_{fouling} and λ_{fouling} are, respectively, the thickness and thermal conductivity of the fouling layer.

3.3.2 Expressions for Convective Heat Transfer Coefficients

Given that convection involves heat exchange caused by fluid flow, it is essential to understand flow characteristics, particularly in the vicinity of the area where heat exchange occurs. It is however unfortunate that flows close to interfaces, especially under non-isothermal conditions, cannot be characterised easily. Empirical models have therefore been developed, which relate convective heat transfer coefficients with flow, physical and thermal properties of fluids. These models are normally expressed in the form of equations containing dimensionless numbers. For forced convection, the four common dimensionless numbers used are the Nusselt number (Nu), Reynolds number (Re), Prandtl number (Pr) and Stanton number (St). Nusselt and Reynolds numbers contain geometric length scale, and therefore their definition can vary with the geometry of the system under consideration. Since convective heat transfer in cylindrical pipes is common, the dimensionless numbers for this case is considered below:

$$\text{Nusselt number : Nu} = \frac{hd}{\lambda}; \qquad (3.27)$$

$$\text{Reynolds number, Re} = \frac{du\rho}{\mu} = \frac{du}{\upsilon}; \qquad (3.28)$$

$$\text{Prandtl number, Pr} = \frac{C_P\mu}{\lambda} = \left(\frac{\mu}{\rho}\right)\left(\frac{\rho C_P}{\lambda}\right) = \frac{\upsilon}{\alpha}; \text{ and} \tag{3.29}$$

$$\text{Stanton number, St} = \frac{h}{\rho\upsilon C_P} = \frac{\text{Nu}}{\text{Re.Pr}} \tag{3.30}$$

where h is the heat transfer coefficient; d is the pipe diameter; λ is the fluid thermal conductivity; C_P is the fluid specific heat; μ is the fluid viscosity; $\nu = \mu/\rho$ is the fluid kinematic viscosity; and $\alpha = \lambda/\rho C_P$ is the thermal diffusivity. Given that St is directly related to Nu, Pr and Re, no new information is obtained by introducing it, except algebraic convenience since it crops up in quite naturally in theoretical analyses of convection. The Nusselt number gives the heat transfer coefficient, whereas the Reynolds number states whether the flow is laminar or turbulent. The Prandtl number, as evident above, compares momentum diffusivity (ν) with thermal diffusivity (α).

Since natural convection involves flow under gravity induced by density gradients, expressions for heat transfer involve another dimensionless number known as the Grashof number, defined as:

$$\text{Gr} = \frac{\beta g\rho^2 L^3 \Delta T}{\mu^2} = \frac{\beta g L^3 \Delta T}{\upsilon^2} = \frac{\text{buoyancy force}}{\text{viscous force}} \tag{3.31}$$

Here β is the coefficient of thermal expansion which characterises the change in density causing flow; L is the characteristic length scale depending on the geometric configuration of the system under consideration; and ΔT is the temperature difference, typically between the heat transfer surface and the bulk. In general, β is defined as follows:

$$\beta = -\frac{1}{\rho}\left(\frac{\partial\rho}{\partial T}\right)_p \tag{3.32}$$

For an ideal gas, it can be shown that $\beta = 1/T$. For natural convection, the Rayleigh number, which is the product of Grashof and Prandtl numbers, is also widely used in correlations.

It may be noted that, in addition to cylindrical pipes, (1) *plate heat exchangers* – which are made of stacks of flat plates with hot and cold fluids flowing alternately between the plates and (2) *scraped surface heat exchangers*, which are well suited for highly viscous systems where the product is brought into contact with a heat transfer surface that is rapidly and continuously scraped, in order to expose the surface to the passage of untreated product, are extensively used in food processing operations.

Empirical equations relating these dimensionless numbers for a variety of situations are listed in Table 3.3.

Table 3.3 Expressions used to estimate heat transfer coefficients

Turbulent flow through pipes (Dittus-Boelter equation)	$Nu = 0.023Re^{0.8}Pr^{0.4}$	$Re > 5000$
Laminar flow through pipes (Sieder-Tate equation)	$Nu = 1.86(Re . Pr . d/L)^{1/3}(\mu/\mu_w)^{0.14}$	d/L is the ratio of the pipe diameter to length; (μ/μ_w) is the ratio of the fluid viscosity at the centre of the pipe to that at the wall.
Flow over a flat plate (Forced convection)	$Nu = 0.66Re^{1/2}Pr^{1/3}$ $Nu = 0.036Re^{0.8}Pr^{1/3}$	$Re < 10^5$ $Re > 10^5$ the characteristic length scale in Re and Nu is the length of the plate measured along the flow
Flow over a horizontal flat plate (Natural convection, hot side facing up)	$Nu = 0.54Ra^{0.25}$ $Nu = 0.14Ra^{1/3}$ $Ra = $ Rayleigh number $= Gr \times Pr$	$10^5 < Ra < 2 \times 10^7$ $2 \times 10^7 < Ra < 3 \times 10^{10}$ the characteristic length scale is the length of the plate measured along the flow
Flow over a horizontal flat plate (Natural convection, hot side facing up)	$Nu = 0.27Ra^{0.25}$ $Ra = $ Rayleigh number $= Gr \times Pr$	$3 \times 10^5 < Ra < 10^{10}$ the characteristic length scale is the length of the plate measured along the flow
Flow over a heated vertical surface (Natural convection)	$Nu = \left(0.825 + \dfrac{0.39\left(Ra^{1/6}\right)}{\left[1+(0.49/Pr)^{9/16}\right]^{8/27}}\right)^2$ $Ra = $ Rayleigh number $= Gr \times Pr$	The characteristic length is the height of the surface
Flow over a horizontal cylinder (Forced convection)	$Nu = BRe^nPr^{1/3}$	$B = 0.99, n = 0.33, 0.4 < Re < 4$ $B = 0.91, n = 0.38, 4 < Re < 40$ $B = 0.68, n = 0.37,$ $40 < Re < 4000$ $B = 0.19, n = 0.62,$ $4000 < Re < 40000$ $B = 0.03, n = 0.80, Re > 40000$ the characteristic length scale in Re and Nu is the cylinder diameter
Flow over a short vertical cylinder of height L (Natural convection)	$Nu = \left(0.825 + \dfrac{0.39\left(Ra^{1/6}\right)}{\left[1+(0.49/Pr)^{9/16}\right]^{8/27}}\right)^2$ $Ra = $ Rayleigh number $= Gr \times Pr$	The expression is the same as for natural convection over a heated vertical surface. The characteristic length is the cylinder height (L); further: $D/L \geq 35/Gr^{0.25}$
Flow over a horizontal cylinder (Natural convection)	$Nu = \left(0.6 + \dfrac{0.39\left(Ra^{1/6}\right)}{\left[1+(0.56/Pr)^{9/16}\right]^{8/27}}\right)^2$ $Ra = $ Rayleigh number $= Gr \times Pr$	$10^{-5} < Ra < 10^{12}$ The characteristic length scale is the cylinder diameter
Flow over sphere (Natural convection)	$Nu = 2 + 0.43Ra^{0.25}$ $Ra = $ Rayleigh number $= Gr \times Pr$	$1 < Ra < 10^5, Pr \approx 1$
Flow over sphere (Forced convection)	$Nu = 2 + (0.4Re^{1/2} + 0.06Re^{2/3})Pr^{0.4}$	$3.5 < Re < 7.6 \times 10^4$ and $0.71 < Pr < 380$

It has to be stressed that the equations have been determined by fitting experimental data, as a consequence of which uncertainty in the magnitude of Nu can be as high as 30%

Problem 3.2 Orange juice, flowing at 1 kg s^{-1} is to be pasteurised by heating from 10 °C to 80 °C in a shell and tube heat exchanger. Water enters at 90 °C and flows counter-currently to the juice, leaving the heat exchanger at 34 °C. The heat exchanger consists of tubes having a length of 1.50 m and diameter of 0.026 m. If the overall heat transfer coefficient is 1700 W m^{-2} K^{-1}, determine the necessary mass flow rate of water and the number of tubes required. The mean heat capacity of orange juice may be assumed to be 3.80 kJ kg^{-1} K^{-1} while that of water may be assumed to be 4.18 kJ kg^{-1} K^{-1}.

It would be best to understand this problem by relating the flows involved to Fig. 3.5b, where $T_{mi} = 10$ °C, $T_{mo} = 80$ °C, $T_{wi} = 90$ °C and T_{wo} needs to be determined by undertaking a heat balance, i.e. by assuming that the rate at which the heat is gained by the orange juice is equal to the rate at which the heat is lost by hot water. The rate of heat gain by the flowing orange juice, $Q =$ (mass flow rate) (Specific heat of the orange juice)$(T_{mo}-T_{mi}) = m\ C_p\ \Delta T. = 1 \times 3.8$ (80–10) = 266 kW. Likewise, if W is the mass flow rate of water in kg s^{-1}, the rate of heat transfer is also $Q = W \times 4.18 \times (90{-}34)$ whence $W = 1.14$ kg s^{-1}.

The temp differences at the two ends of the counter-current flows is: (90–80)K and (34–10)K, giving a log mean temp difference of 16 K based on Eq. 3.24. Since, according to Eq. 3.25, $Q = U A \Delta T_{ln}$, and $\underline{U} = 1700$ W m^{-2} K^{-1}, it follows that the total heat transfer area $A = 9.78$ m^2. Given the length of each tube is 1.5 m and its diameter is 0.026 m, the heat transfer area of each tube is $(\pi \times 0.026 \times 1.5)$ m^2, and the number of tubes will be given by $9.78/(\pi \times 0.026 \times 1.5) = 80$ (approximately).

Problem 3.3 Water flows at 80 °C with a mass flow rate of 12 kg s^{-1} through a thin-walled circular pipe of 50 mm internal diameter and length of 20 m, which is surrounded by air at 20 °C. The heat transfer coefficient at the outer surface of the pipe is 200 Wm^{-2} K^{-1}, whereas, at the inner surface, the heat transfer coefficient is given by the Dittus-Boelter correlation: Nu = 0.023 Re$^{0.8}$Pr$^{0.4}$. Estimate the overall heat transfer coefficient, and hence the heat flux (i.e. heat transfer rate per unit surface area) through the pipe.

Data: For flow through the pipe: Reynolds number Re = $du\rho/\mu$; Nusselt number Nu = $h_i d/k$; and Prandtl number Pr = $C_p\mu/k$, where d is the pipe diameter; u is the mean fluid velocity; and h_i is the internal heat transfer coefficient. For water: density $\rho = 1000$ kgm^{-3}, viscosity $\mu = 8 \times 10^{-4}$ Pas, specific heat $C_p = 4200$ J kg^{-1}K^{-1}, and thermal conductivity $k = 0.5$ Wm^{-1} K^{-1}.

During use, a fouling scale of conductivity 0.2 Wm^{-1} K^{-1} gets deposited on the inner tube wall. If the thickness of the scale is 0.5 mm, calculate the reduced heat flux.

The pipe diameter $d = 0.05$ m, therefore the cross sectional area of the pipe = $(\pi/4)\ d^2 = 1.96 \times 10^{-4}$ m^2. Since the mass flow rate of the liquid is 12 kg s^{-1}, and its density is 1000 kg m^{-3}, the volumetric flow rate will be 0.012 m^3 s^{-1}, whence, its velocity $u = 0.012/[(\pi/4)\ d^2] = 6.11$ m s^{-1}. Also Re = $du\rho/\mu = 0.05 \times 6.11 \times 1000/(8 \times 10^{-4}) = 381875$. and the Prandtl number is given by Pr = $C_p\mu/$

$k = 4200 \times 8 \times 10^{-4}/0.5 = 6.72$. Hence, employing the Dittus-Boelter correlation given in the statement of this problem, $h_i = 14395 \ \text{Wm}^{-2}\,\text{K}^{-1}$. It is also known from the statement of this problem that $h_o = 200 \ \text{Wm}^{-2}\,\text{K}^{-1}$. Hence the overall heat transfer coefficient, U can be obtained by employing Eq. 3.23 for the case where the pipe is assumed to be very thin, i.e. $1/U = 1/h_i + 1/h_o$ to yield $U = 197.26 \ \text{Wm}^{-2}\,\text{K}^{-1}$. Given that the flux is the rate of heat transfer per unit area of the heat transfer surface (in this case, the area of the pipe exposed to the ambient), flux $= U\Delta T = 197.26\,(80\text{--}20) = 11836 \ \text{W m}^{-2}$.

If a fouling scale forms on the inner surface of the tube, i twill offer an additional layer of resistance to heat transfer given by (x/λ) where x is the thickness of the fouling layer and λ is its thermal conductivity, which will work out to $(0.0005/0.2) = 0.0025 \ \text{W}^{-1}\text{m}^2\text{K}$. By adding this resistance in Eq. (3.23), the overall heat transfer coefficient $1/U = 1/h_i + 1/h_o + (x/\lambda)$ giving $U = 132.11 \ \text{Wm}^{-2}\,\text{K}^{-1}$, and a heat flux of $7926.6 \ \text{W m}^{-2}$. It may be noted that a significant assumption has been made in this calculation, i.e. the formation of the scale does not influence the temperature of the pipe wall – which is not the case. A much more rigorous calculation has to be undertaken to determine the various temperature driving forces and the flux.

3.4 Radiative Heat Transfer

The modes of heat transfer covered above, i.e. conduction and convection, involve movement of heat through material media. Radiative heat, on the other hand, does not require a material medium for propagation. It is a component of the spectrum of electromagnetic energy emitted by molecules which have been excited by temperature. Electromagnetic radiations differ in wavelength (and hence in frequency), but all radiations travel at the speed of light in free space ($\approx 3 \times 10^8 \ \text{ms}^{-1}$). Radiations in the infra red region of the spectrum (see Fig. 3.6) can be directly absorbed by matter in the form of sensible heat. Radiation in other components of the spectrum shown in Fig. 3.6, notably in the radio and microwave region, when absorbed by matter, can be converted by the absorbing material into heat energy depending on the dielectric properties* of the material. This is indeed the principle underpinning *dielectric* and

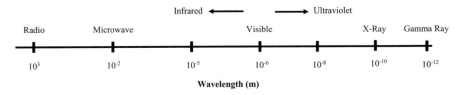

Fig. 3.6 Spectrum of electromagnetic radiations. The figure is not drawn to scale, and it mainly serves to illustrate the relative positions of commonly acknowledged radiations. The frequency of each radiation can simply be obtained by dividing the speed of light $c = 3 \times 10^8 \ \text{ms}^{-1}$ by the wavelength in metres

microwave heating of foods. Radiation is much more sensitive to temperature and it is of dominating importance in the operation of ovens and freeze dryers.

The quantitative description of thermal radiation can be best keyed to the radiation characteristics of the "ideal radiator" known as the *blackbody*. There are two characteristic features of a blackbody: (i) it absorbs all the radiation incident on its surface; and (ii) the quality and intensity of the radiation it emits are solely determined by its absolute temperature. The *Stefan-Boltzmann law* describes the radiative flux from a blackbody as:

$$q = \sigma T^4 \tag{3.33}$$

This flux is uniform in space, in other words, distributed over a hemisphere. The temperature T is the absolute temperature, i.e. expressed in K; and the Stefan-Boltzmann constant, σ, takes a value of 5.67×10^{-8} Wm^{-2} K^{-4}.

The spectral distribution of energy flux from a blackbody is expressed in terms of the product $E_\lambda d\lambda$, which represents the hemispherical flux density lying in the wavelength range λ and $(\lambda + d\lambda)$. According to *Planck's law*:

$$E_\lambda d\lambda = \frac{n^2 c_1 \lambda^5}{e^{c_2/\lambda T} - 1} d\lambda \quad \text{or} \tag{3.34}$$

$$\frac{E_\lambda}{n^2 T^5} = \frac{c_1 (\lambda T)^{-5}}{e^{c_2/\lambda T} - 1} \tag{3.35}$$

In the above equations, n represents the refractive index of the emitter; and c_1 and c_2 are the first and second Planck's law constants, taking values 3.74×10^{-16} Jm^{2}s^{-1} and 1.44×10^{-2} m K, respectively. It may be noted that the Stefan Boltzmann law (Eq. 3.33) is essentially an integral of Planck's equation (Eq. 3.34) with respect to λ, between $\lambda = 0$ and ∞. In other words, the total black body emission per unit area per unit time is:

$$\int_0^\infty E_\lambda d\lambda = \int_0^\infty \frac{n^2 c_1 \lambda^5}{e^{c_2/\lambda T} - 1} d\lambda = q = \sigma T^4 \tag{3.36}$$

It is evident from Eq. (3.35) that $(E_\lambda / n^2 T^5)$ is a function of the product (λT) alone. It can also be shown that $(E_\lambda / n^2 T^5)$ attains a maximum value when $\lambda T \approx 2.9 \times 10^{-3}$ m K. Thus, the wavelength of maximum emissive power is inversely proportional to the absolute temperature; this is known as the *Wien's displacement law*.

In general, the radiative flux emitted by any body is given by:

$$q = \varepsilon \sigma T^4 \tag{3.37}$$

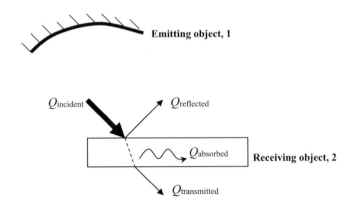

Fig. 3.7 Interaction between a surface and incident radiation

In the above equation, ε is a material property called *emissivity*, which assumes a value between zero and 1; for a black body, $\varepsilon = 1$. It is clear from a comparison of Eqs. (3.33) and (3.37) that emissivity represents the ratio of the radiation emitted by a surface to the radiation emitted by a perfect emitter, i.e. a blackbody, at the same temperature. It is a measure of how efficiently a surface emits radiation.

When emitted radiation strikes a second surface, it is reflected, absorbed, and transmitted (Fig. 3.7); the extent to which each occurs depends on the material. The component that contributes to the heating of the material is the absorbed radiation. The amount of heat absorbed by object 2 in Fig. 3.7 is given by:

$$(Q_{absorbed})_2 = \alpha_2 (Q_{incident})_2 \tag{3.38}$$

where the fraction of the incident radiation that is absorbed, α_2, is called the absorptivity. The incident radiation is determined by the amount of radiation emitted by the original object and how much of the emitted radiation actually strikes the second surface. The latter is given by the *shape factor* or the *view factor*, $F_{1 \to 2}$, which is a measure of the radiation emitted by object 1 that reaches the surface of object 2.

$$(Q_{incident})_2 = F_{1 \to 2}(Q_{emitted})_1 = F_{1 \to 2}\varepsilon_1 \sigma A_1 T_1^4 \tag{3.39}$$

Combining Eqs. (3.38) and (3.39), the expression for the amount of heat absorbed by object 2 is:

$$(Q_{absorbed})_2 = \alpha_2 F_{1 \to 2}\varepsilon_1 \sigma A_1 T_1^4 \tag{3.40}$$

It is evident that object 2 will also radiate or emit heat depending on its emissivity (ε_2), area (A_2) and temperature (T_2); i.e.:

$$(Q_{emitted})_2 = \varepsilon_2 \sigma A_2 T_2^4 \tag{3.41}$$

The net radiative heat transfer rate at the surface of object 2 is therefore given by:

$$Q_2 = (Q_{absorbed})_2 - (Q_{emitted})_2 = \alpha_2 F_{1\rightarrow2} \varepsilon_1 \sigma A_1 T_1^4 - \varepsilon_2 \sigma A_2 T_2^4 \tag{3.42}$$

Likewise, the net heat transfer rate at the surface of object 1 is given by:

$$Q_1 = \alpha_1 F_{2\rightarrow1} \varepsilon_2 \sigma A_2 T_2^4 - \varepsilon_1 \sigma A_1 T_1^4 \tag{3.43}$$

The shape or view factor, F, in Eqs. (3.42) and (3.43), is a geometric factor which, in general, is determined by the shapes and relative locations of two surfaces. For two surfaces, 1 and 2, it can be shown that:

$$F_{1\rightarrow2} = F_{2\rightarrow1} \left(\frac{A_2}{A_1} \right) \tag{3.44}$$

Tabulated values of F for different situations are available in published literature (e.g. Incropera, F.P. and De Witt, D.P., *Introduction to Heat Transfer, Second Edition*, John Wiley & Sons, New York, NY, 1990).

If object 1 is small and object 2 is enclosing it, all the heat emitted by object 1 will reach the surface of object 2, i.e.: $F_{1\rightarrow2} = 1$; and $F_{2\rightarrow1} = A_1/A_2$ from Eq. (3.44). By dividing both sides of Eq. (3.43) by A_1 and substituting for $F_{2\rightarrow1}$, the net heat transfer rate is given by:

$$q_1 = \frac{Q_1}{A_1} = \alpha_1 \varepsilon_2 \sigma T_2^4 - \varepsilon_1 \sigma T_1^4 \tag{3.45}$$

If object 1 is a *grey body*, i.e. a body having emissivity and absorptivity equal and independent of wavelength, $\alpha_1 = \varepsilon_1$; and object 2 is a black body, $\varepsilon_2 = 1$, Eq. (3.45) becomes:

$$q_1 = \varepsilon_1 \sigma \left(T_2^4 - T_1^4 \right) \tag{3.46}$$

In general, the right hand side of Eq. (3.46) can be $\left(T_2^4 - T_1^4 \right)$ or $\left(T_1^4 - T_2^4 \right)$ depending on whether T_2 or T_1 is greater.

3.5 Selected Practical Aspects of Heat Transfer in Food Systems

3.5.1 Transient Heat Transfer Between a Solid and Surrounding Liquid

The term transient refers to time-dependent heat transfer. In other words, we are considering a non-equilibrium or unsteady state process, similar to the conduction problems described in the discussion leading to Eq. (3.16). Examples of transient heat transfer between a solid and surrounding fluid includes cooking of lumps of meat or vegetables, and sterilising particles suspended in a sauce inside a can. In these processes, the heating medium transfers the heat to the liquid phase which is either in a state of natural or forced convection. The surface of the solid particle, which is at a lower temperature than the liquid, receives heat by *convection*, and *conducts* it radially towards its centre. Thus, the normal temperature gradient in this process is as shown in Fig. 3.8.

There are two resistances to heat transfer: i) an external convective resistance, which is proportional to $1/h$ where h is the liquid-side heat transfer coefficient (see Sect. 3.3.1) and ii) an intra-particle conductive resistance which is proportional to R/λ where R is the particle length scale (e.g. radius, if the particle is spherical) and λ is its thermal conductivity. The ratio of the two resistances, expressed as conductive to convective resistances, hR/λ, is known as the *Biot number* (*Bi*). This dimensionless number indicates which resistance dominates. For instance, a low value of Biot number (<<1) indicates that convective resistance dominates. We can therefore discuss transient heat transfer between the liquid and the particle for two cases: low and high values of *Bi*. When *Bi* is very low, say <0.1, the particle resistance is very low and the temperature inside it is uniform, but lower than the outside temperature; the difference between the two temperatures provides the driving force for heat transfer. Furthermore, if the volume of the surrounding fluid is large enough so that its temperature remains a constant throughout the process, a simple model can be derived to represent heat transfer. In this model, T_L remains constant and $T_S = T_C = T(t)$. The rate of heat transfer from the liquid at any instant is $hA(T_L -$

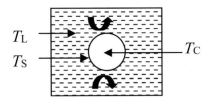

Fig. 3.8 Heat transfer between a solid particle and surrounding liquid. T_L, T_S and T_S are the temperatures of the liquid (assumed to be uniform through out), particle surface and the particle centre, respectively. Furthermore, $T_L > T_S > T_C$. Normally, the particle is assumed to be completely heat treated, when T_C attains the heat treatment temperature and it is held at that temperature for a stipulated period of time

$T(t)$) where A is the heat transfer area; this should be equal to the rate at which the sensible heat of the particle increases, i.e.,

$$hA(T_L - T(t)) = \frac{d}{dt}\left(mC_pT(t)\right) \tag{3.47}$$

where m is the particle mass and C_p is its specific heat. If the initial temperature of the particle is $T(0)$, the above equation can be solved to yield the following expression for the particle temperature at any time:

$$\frac{T_L - T(t)}{T_L - T(0)} = \exp\left(-\frac{hA}{mC_P}t\right) \tag{3.48}$$

If the external convective resistance is low in relation to the intra-particle conductive resistance, the temperatures T_L and T_S (see Fig. 3.8) will not be too different. In practice, this can be achieved by ensuring that the liquid is sufficiently well agitated. The internal resistance, on the other hand, will create a temperature gradient inside the particle running from the surface to its centre. It should also be noted that the temperature gradients vary with time, measured from the instant the particle is brought into contact with the fluid. Typical temperature profiles are shown in Fig. 3.9. If the convective resistance is negligible, the particle surface temperature will be constant and equal to T_L at all times. This situation decomposes to being a problem on pure conduction, i.e. the situation leading to Eq. (3.17) which yields the temperature distribution within the particle at any time (note that T_A in Eq. (3.17) will be replaced by T_L and x will represent the radial position). In other words,

$$\frac{T(r,t) - T_L}{T(0) - T_L} = \mathrm{erf}\left(\frac{r}{2\sqrt{\alpha t}}\right) \tag{3.49}$$

If both external and intra-particular resistances are comparable, the following equation gives transient values of temperature inside the particle at any position:

$$\frac{T(r,t) - T(0)}{T_L - T(0)} = \mathrm{erfc}\left(\frac{r}{2\sqrt{\alpha t}}\right) - \left[\exp\left(\frac{hr}{\lambda} + \frac{h^2\alpha t}{\lambda^2}\right)\right]\left[\mathrm{erfc}\left(\frac{r}{2\sqrt{\alpha t}} + \frac{h\sqrt{\alpha t}}{\lambda}\right)\right]$$

The values of the complementary error function $\mathrm{erfc}(x) = 1 - \mathrm{erf}(x)$ are given in Table 3.2.

In general, it is very difficult to determine transient temperatures in particles having various shapes when, both, external and intra-particular resistances to heat transfer are significant. Digital and graphical solutions involving the Biot number ($\mathrm{Bi} = hL/\lambda$), Fourier number ($\mathrm{Fo} = \alpha t/L^2$) and normalized initial centerline temperature $\theta_C = \frac{T_C(t) - T_L}{T(0) - T_L}$ are available. We will consider here the graphical solutions in the form of charts, because they are more illustrative. The solution is in the form of charts called Heisler Charts. There are two types of charts available for each

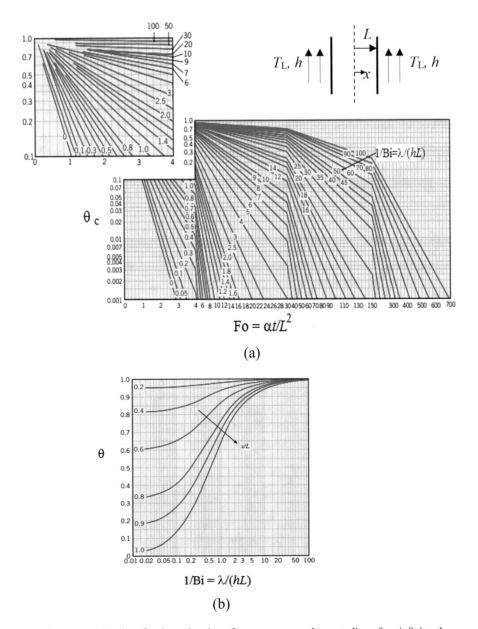

Fig. 3.9 (a) Heisler chart for the estimation of temperatures at the centerline of an infinite plane $T_C(t)$ at any time t when both external and internal heat transfer resistances are in play. The dimensionless centerline temperature is given on the Y-axis as: $\theta_C = \frac{T_C(t)-T_L}{T(0)-T_L}$; note that L is half the slab thickness; T_L is the constant bulk temperature surrounding the plane; $T(0)$ is the uniform initial temperature of the plane; h is the external heat transfer coefficient; λ is the thermal conductivity of the material; 1/Bi is the reciprocal Biot number; Fo is the Fourier number defined on the X axis, where α is the thermal diffusivity of the material of the plane. (b) Heisler chart for estimating the transient temperature along any line located at a distance x from the centerline $T(x,t)$. The dimensionless temperature on the Y axis is given by $\theta = \frac{T(x,t)-T_L}{T(0)-T_L}$. (Reproduced with Permission: M.P.Heisler, Transactions ASME **69**, 227–236 (1947)

geometry: the first for finding the centreline temperature, and the second for finding the temperature at any location; see Figs. 3.9, 3.10, and 3.11. The first chart has the Fourier number, Fo, on the horizontal axis, θ_C on the vertical axis, and lines representing different values of the inverse of Biot number, i.e. Bi^{-1}. When the temperature at a given position and time, inside the particle of known shape, is to be determined, Fo and Bi^{-1} are first calculated and θ_C is read off this chart. To find the temperature at any location within the particle, the second chart is used. This chart has Bi^{-1} on the horizontal axis, normalized temperature (θ/θ_C) on the vertical axis, and lines corresponding to normalized position, x/L for plane or r/r_0 for sphere and cylinder. From the values of Bi^{-1} and normalized position, θ/θ_C can be read off this graph, and knowing θ_C the value of θ can be determined. Finally, noting that $\theta = \frac{T(x,t)-T_L}{T_C(t)-T_L}$, $T(x.t)$ can be estimated.

Problem 3.4 **A spherical meat ball, 50 mm in diameter, is taken out of a freezer where it is uniformly frozen at -18 °C, and directly placed in a convection oven pre-heated to 120 °C, where the convective heat transfer coefficient is 120 Wm^{-2} K^{-1}. Calculate the temperature at the centre of the meatball, and at the radius of 5 mm, 30s after placing it in the oven. The thermal conductivity of the meat ball is 2.2 Wm^{-1} K^{-1}; its specific heat is 2050 J kg^{-1} K^{-1} and density is 920 kg m^{-3}.**

In this problem $r_0 = 25$ mm or 0.025 m; $\lambda = 2.2$ Wm^{-1} K^{-1}; $C_p = 2050$ J kg^{-1} K^{-1}; $\rho = 920$ kg m^{-3}; and $h = 120$ Wm^{-2} K^{-1}. Thus, the thermal diffusivity of the meat ball given by $\alpha = \lambda/(\rho C_p) = 1.16 \times 10^{-6}$ m^{2}s^{-1}. With Froude number Fo $= \alpha t/r_o^2 = 0.056$ and the reciprocal of the Biot number $Bi^{-1} = \lambda/(hr_0) = 0.73$, Fig. 3.11a yields the dimensionless centre temperature as follows:

$$\theta_C = \frac{T_C(t) - T_L}{T(0) - T_L} = \frac{T_C(t) - 120}{-18 - 120} = \frac{T_C(t) - 120}{-138} = 0.45,$$

whence the centre temperature, 30s after placing in the oven is: $T_C(30s) = 57.9$ °C. When $r = 5$ mm, $r/r_o = 0.2$. Now referring to Fig. 3.11b, the dimensionless temperature corresponding to $r/r_o = 0.2$ and $Bi^{-1} = 0.73$ is given by:

$$\frac{T_r(30s) - 120}{57.9 - 120} = 0.94 \text{ which yields } T_r(30s) = 61.6°C$$

The assumption that heat transfer rates are determined by pure conduction in the meat ball is a gross oversimplification. The frozen meat ball has ice which melts as it gets heated, so there can be some convective movements inside the meat ball which can enhance the heat transfer rates more than what we estimate by pure conduction. Moreover, even if pure conduction dominates heat transfer, the physical properties need not be constant as the temperature rises. All these phenomena can be taken into account in order to develop a significantly more complex model which is likely to be more realistic.

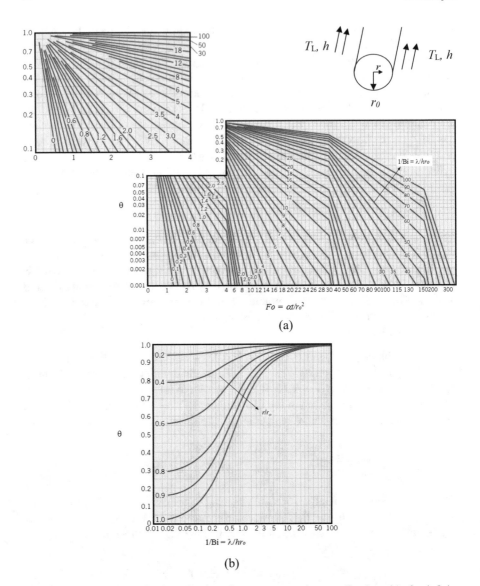

$$Fo = \alpha t/r_0^2$$

(a)

(b)

Fig. 3.10 (**a**) Heisler chart for the estimation of temperatures at the centerline (or axis) of an infinite cylinder $T_C(t)$ at any time t when both external and internal heat transfer resistances are in play. The dimensionless centerline temperature is given on the Y-axis as: $\theta_C = \frac{T_C(t)-T_L}{T(0)-T_L}$; note that r_0 is the radius of the infinite cylinder; T_L is the constant bulk temperature surrounding the cylinder; $T(0)$ is the uniform initial temperature of the cylinder; h is the external heat transfer coefficient; λ is the thermal conductivity of the material; 1/Bi is the reciprocal Biot number; Fo is the Fourier number defined on the X axis, where α is the thermal diffusivity of the material of the cylinder. (**b**) Heisler chart for estimating the transient temperature along any line located at a radial distance r from the centerline T(r,t). The dimensionless temperature on the Y axis is given by $\theta = \frac{T(r,t)-T_L}{T(0)-T_L}$. (Taken from https://en.wikipedia.org/wiki/Heisler_chart)

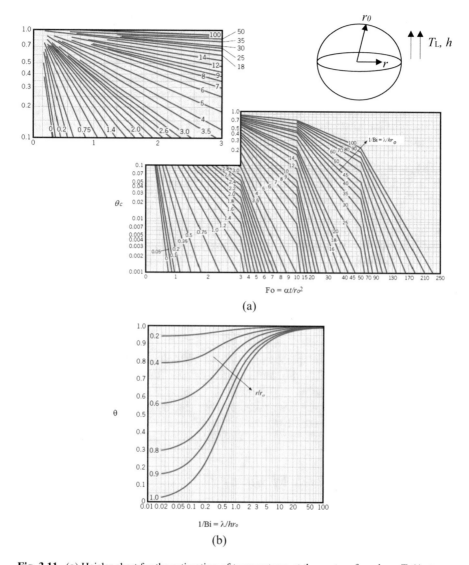

Fig. 3.11 (a) Heisler chart for the estimation of temperatures at the center of a sphere $T_C(t)$ at any time t when both external and internal heat transfer resistances are in play. The dimensionless centerline temperature is given on the Y-axis as: $\theta_C = \frac{T_C(t)-T_L}{T(0)-T_L}$; note that r_0 is the radius of the sphere; T_L is the constant bulk temperature surrounding the sphere; $T(0)$ is the uniform initial temperature of the sphere; h is the external heat transfer coefficient; λ is the thermal conductivity of the material; 1/Bi is the reciprocal Biot number; Fo is the Fourier number defined on the X axis, where α is the thermal diffusivity of the material of the sphere. (b) Heisler chart for estimating the transient temperature at a radial distance r from the center of the sphere $T(r,t)$. The dimensionless temperature on the Y axis is given by $\theta = \frac{T(r,t)-T_L}{T(0)-T_L}$. (Reproduced with Permission M.P.Heisler, Transactions ASME **69**, 227–236 (1947)

Problem 3.5 A suspension of starch particles (0.8 mm in diameter) in water, weighing 500 kg, is to be heated from 10 °C to 85 °C in a steam jacketed stirred vessel where the heat transfer coefficient is 200 Wm^{-2} K^{-1}. The steam is at a temperature of 120 °C and it transfers heat to the suspension by condensing at this temperature. The area for heat transfer between the steam and the suspension is 3.8 m^2, while the steam side heat transfer coefficient is 8000 Wm^{-2} K^{-1}. The thermal conductivity of starch particles is 0.4 Wm^{-1} K^{-1}; and the specific heat of the suspension is 3.9 kJ kg^{-1} K^{-1}. How long will it take for the suspension to reach 85 °C?

The particle Biot number is given by: Bi $= hr_0/\lambda = 0.2$ which is significantly less than 1 and suggests that the heat transfer coefficient external to the particle is controlling the rate of heat transfer. In other words, the particle temperature at any time will be uniform and the same as the liquid temperature. Furthermore the steam side heat transfer coefficient is significantly greater than the suspension side heat transfer coefficient, so the steam side thermal resistance can be safely neglected. Thus, Eq. 3.48 can be used to estimate the time taken for the suspension to heat up, i.e.

$$\frac{T_L - T(t)}{T_L - T(0)} = \exp\left(-\frac{hA}{mC_P}t\right)$$

Given that T_L, the steam temperature $= 120$ °C, $T(0)$, the initial suspension temperature $= 10$ °C, and $T(t)$ the temperature of the suspension at the end of heating $= 85$ °C, with $h = 200$ Wm^{-2} K^{-1}, $A = 3.8$ m^2, $m = 500$ kg and $C_p = 3900$ J kg^{-1} K^{-1}, the time taken to heat can be estimated from the above equation to be 2938 s (or 0.82 h).

3.5.2 Heating of Foods by Microwaves

It was noted in Sect. 3.4 that certain radiations, when absorbed by matter, can be converted into heat energy. Microwave radiation refers to the region of the radiation spectrum with frequencies between $\sim10^9$ Hz to $\sim10^{11}$ Hz. This type of radiation lies between infrared radiation and radio waves. Although the frequency range of microwaves covers two orders of magnitude, the food industry is only allowed to use very specific frequencies, so that there is no interference with other applications such as telecommunications. The frequency of radiation used in microwave ovens is approximately 2.5×10^9 Hz. We also know that the product of the frequency and wavelength of any radiation must be equal to the speed of light (i.e. $c = \nu\lambda$). Assuming $c = 3 \times 10^8$ ms^{-1}, the wavelength of microwave works out to be 0.12 m which is a very low energy radiation. Indeed, the microwave radiation used to heat food is far less energetic than visible light. This naturally begs the question, why are microwaves so efficient at heating foods? The answer lies in the

Fig. 3.12 Flipping of
dipoles in water molecule
with phase change of
incident
electromagnetic wave

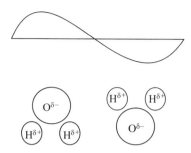

way these waves interact with the water molecules present in foods. Given the difference between the electro-negativities of oxygen and hydrogen, water molecules act as dipoles with a partial positive charge on hydrogen atoms and a partial negative charge on oxygen. It is well known that dipoles tend to orient themselves to applied electric fields depending on the nature of the charge on the electric field. Within the span of a single cycle, the oscillating electromagnetic radiation changes its electrical nature twice as shown in Fig. 3.12. This flips the orientation of water molecules twice. In other words, a 2.5×10^9 Hz microwave will flip water molecules 5×10^9 times or 5 billion times in one second! This vigorous movement of water molecules raises its temperature and that of the food as a whole.

Most foods contain over 70% water by mass, making microwaves an effective way for heating. Microwave radiations also have the ability to penetrate several inches, and radiations with longer wavelengths tend to penetrate deeper. However, there is a limit. If the food were larger than several inches thick, the middle of the food can only receive heat by conduction or convection from the parts heated by microwaves. The negative side of using microwaves is that food with low water contents take longer to heat. Furthermore, frozen foods also take longer to heat because of the restricted movement of the water molecules. It is also necessary to note that foods heated in a microwave cannot become hotter than the boiling point of water; hence foods cannot be browned in a microwave and crusts cannot be formed. Another major drawback of microwave is the formation of hot spots. In practice, microwave radiation is made to undergo multiple reflections inside the cooking chamber, which causes radiations to interact with other reflected radiation in such way that hot and cold spots are formed inside the microwave. The phenomenon responsible for this is the interference of waves, i.e. waves overlapping in such way that the crests match one another interfere constructively, to form a hot spot. Waves which interfere destructively result in a cold spot. The uneven heating caused by these effects can be reduced significantly by using a rotary platform, which is a common feature in many models of microwave ovens.

Microwave propagation in foods depends on the dielectric and the magnetic properties of the food. The electromagnetic properties of the food are characterized by complex permittivity (ϵ) and complex permeability (μ). Complex numbers containing real and imaginary parts are used to describe permittivity and permeability due to the wave or cyclic nature of electromagnetic waves. Very broadly speaking, the former represents electrical interaction and the latter represents magnetic interactions. Heating is largely an electrical effect and the permittivity plays the dominant role. The complex permittivity is composed of a real part (ϵ') known as the dielectric constant, and imaginary part (ϵ'') known as the dielectric loss factor. If i represents the imaginary root ($\sqrt{-1}$), the complex permittivity is defined as:

$$\epsilon = \epsilon' - i\,\epsilon'' \tag{3.50}$$

The dielectric constant (ϵ') describes the material's capacity to store electrical energy whereas the dielectric loss factor (ϵ'') describes the material's ability to dissipate electrical energy as heat. The ratio (ϵ''/ϵ') = tan δ is known as the loss tangent and is a key parameter in the microwave processing of foods. The loss tangent represents a food material's capacity to be penetrated by electrical field and subsequently be dissipated as heat. If an electromagnetic field strength E (Vm^{-1}) is incident at a frequency ϑ on a food, the average thermal power density converted to heat P (Wm^{-3}) is given by:

$$P = 2\pi\vartheta\epsilon_o\epsilon''\,E^2 \tag{3.51}$$

where ϵ_0 is the free space permittivity (8.854×10^{-12} F/m).

In addition to power, the depth of penetration of microwave is also critical. The microwave energy drops with depth and the penetration depth has different definitions. For design calculations, the depth at which the microwave power density drops to 37% (i.e. the fraction 1/e) of its surface value is considered to be the penetration depth, and it is given by:

$$D = \frac{\lambda\sqrt{\epsilon'}}{2\pi\epsilon''} \tag{3.52}$$

Problem 3.6 A flask filled with water (temperature 50 °C, $\epsilon'' = 5.1$) is placed in a microwave heat chamber and exposed to a mean field strength of 2 kV/m at 2450 MHz. What is the power density dissipated? What is the rate of rise of temperature of the water if its density is 988 kg m^{-3} and specific heat is 4.18 kJ kg^{-1} K^{-1}?

With reference to Eq. (3.51), we have $E = 2000$ Vm^{-1}; $\epsilon'' = 5.1$; $\epsilon_0 = 8.854 \times 10^{-12}$ F/m; the frequency $\vartheta = 2.450 \times 10^9$ Hz. The value of $P = 2780$ kW m^{-3}. The rate of rise of temperature per unit volume will therefore be: $P/(\rho C_p) = 0.67$ Km^{-3} s^{-1}.

3.5.3   Ohmic Heating of Foods

Ohmic heating involves making a current flow through the food by placing two electrodes connected to an electrical power supply, normally an alternating current supply (i.e. AC). The food acts as an electrical resistance, and the energy dissipated is converted into heat which raises the temperature of the food. This type of heating is also known as electrical resistance heating or Joule heating. It is interesting to note that the multi-phase nature of the food (i.e. presence of solid and liquid phases together) does not pose any limitation to the flow of electricity. Indeed, fluids and suspended particulate matter, both, heat at comparable rates, which minimizes non-uniform heating of such mixtures and can a tremendous advantage during the thermal processing of particulate foods.

Figure 3.13 shows a schematic description of how Ohmic heating heats up the food. Similar to Eq. 3.51, the average thermal power density P (Wm^{-3}) generated when an electric field strength E (Vm^{-1}) is applied to a food given by:

$$P = \sigma E^2$$

where σ is the electrical conductivity of the material which has the unit of S m^{-1}, i.e. Siemens per meter. It is worth noting that Siemen is dimensionally the reciprocal of Ohm i.e. Ohm^{-1}, also written as mho. In other words, the unit of electrical conductivity is also Ohm^{-1} m^{-1}. The electrical conductivity represents the ease with which the food conducts electricity, but it is not uniform across a food that is being heated. At the same time, the electrical conductivity is also a strong function of temperature and the pH and ionic properties of the food.

The key advantages of Ohmic heating as a method to generate heat in foods are: (1) it is a volumetric method of heating and the efficiency of electrical energy conversion to heat is close to 100%; (2) the chances of fouling are minimized because there are no hot surfaces; (3) the supply of electrical energy can be better

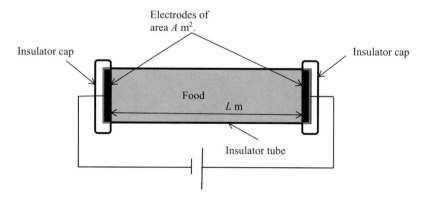

Fig. 3.13 Schematic representation of Ohmic heating of food

controlled; (4) there is no need to mix the food intensely to avoid thermal gradients, so the texture of shear sensitive materials can be protected; and finally, complex multiphase food systems can be heated this way. The key disadvantages are: (1) the electrical energy costs and maximum power available from a local supply, (2) the conductivity limitations of the food material which can be processed by this technology, and (3) the requirement for insulating materials to be used in parts of the equipment, as shown in Fig. 3.13.

References

Carslaw HS, Jaeger JC (1959) Conduction of heat in solids. Oxford University Press
Heisler MP (1947) Trans ASME 69:227–236

Further Reading

Coulson and Richardson's chemical engineering (2017) In: Chhabra RP, Shankar V (eds) Volume 1B heat and mass transfer: fundamentals and applications, 7th edn
Datta AK Biological and bioenvironmental heat and mass transfer. Marcel Dekker Inc. ISBN 0-8247-0775-3
Datta AK, Rakesh V (2013) Principles of microwave combination heating. Compr Rev Food Sci Food Saf 12:24–39
Fryer PJ, de Alwis AAP, Stapley AGF, Zhang L (1993) Ohmic processing of solid-liquid mixtures: heat generation ad convection effects. J Food Eng 18:101–125

Chapter 4
Elements of Mass Transfer

Aim The term mass transfer refers to the movement of molecular species across interfaces. In this chapter, we will consider the driving forces causing mass transfer and the theories underpinning the rates of mass transfer. These theories will help us to understand the design methodology and operation of equipment in which mass transfer occur.

4.1 Introduction

A variety of interfaces are encountered in food processing: interface formed as phase boundaries (solid-liquid, gas-solid, gas-liquid or liquid-liquid) or interface formed by natural or synthetic membranes (e.g. plant cell membrane or cellulose acetate membranes used in ultrafiltration) which are selective to certain species. The movement of species across phase boundaries occurs in a variety of *separation processes*. For instance, during the extraction of coffee, components like sugars, caffein and flavours are transferred from the ground-roast coffee granules into an extraction medium; this is a typical example involving solid-liquid mass transfer. In the production of condensed milk by *evaporation*, water is transferred from milk into a vapour phase; this is an example of vapour-liquid mass transfer. Water is also transferred from grains, fruits and vegetables into air during *drying or de-hydration*, these being examples of mass transfer in gas-solid systems. And finally, the transfer of proteins between aqueous and organic phases, or of components between lipid and aqueous phases, form examples of mass transfer in liquid-liquid systems. In the above examples, the insolubility or poor solubility between the phases resulted in the development of the phase boundary or interface, across which the species moved.

Perfectly miscible fluids can also be separated by selectively permeable walls, known as membranes, which control the movement of molecular species between two miscible fluids. In a large number of cases (not all!), the principles of mass

© Springer Nature Switzerland AG 2022
K. Niranjan, *Engineering Principles for Food Process and Product Realization*,
Food Engineering Series, https://doi.org/10.1007/978-3-031-07570-4_4

transfer are applicable to molecular movement through membranes. For example, the transfer of sugars and flavours from inside plant cells into aqueous media during fruit juice extraction, involve movement across cell membranes. Mass transfer also plays a key role in processes like ultrafiltration where water moves through a cellulose acetate membrane between two aqueous solutions.

It is generally recognized that, just as temperature difference is the cause of heat transfer (see Chap. 3), the difference in *concentrations* of a species between two phases causes the species to transfer from the phase having a higher concentration to the one having a lower concentration. In other words, concentration *gradient* is the driving force for mass transfer between phases. Although, this statement is by and large valid, it is necessary to note that molecular transport in biological systems do not necessarily have to follow this dictum. For example, living cells are known to absorb certain molecular species from extra-cellular solutions where the concentration of the species is *lower* than in the cellular solution. Evidently, there can be other driving forces which facilitate molecular movement against "natural" concentration gradients, but this discussion is beyond the scope of the book.

In this chapter, we will, first of all, consider concentration difference driven movement of a molecular species, which is commonly known as *diffusive mass transfer*. This process does not involve bulk scale movement, and it is analogous to the transfer of heat by *conduction*, already discussed in the Chap. 3. We will then consider the effects of convection, i.e. mass transfer across phases when one or all the phases are subjected to bulk movement, say, by mixing or agitation. The effects of mass transfer on chemical, biochemical and microbial transformations will be discussed in Chap. 5.

4.2 Diffusive Mass Transfer

4.2.1 Fundamentals

Diffusive mass transfer refers to molecular movement solely under the influence of concentration gradient, i.e. the "natural" movement of a solute from a region of high concentration to one with a lower concentration. Concentration gradient, just like temperature gradient in the case of conductive heat transfer (Sect. 3.2), is defined as the slope of the plot of concentration against position (i.e. dC/dx), and has the units: $kg\ m^{-3}/m$. It may be noted that the concentration unit is assumed to be: $kg\ m^{-3}$. The quantitative description of the rate of diffusive mass transfer is given by the Fick's laws. Fick's first law states that the flux of mass transfer (J) i.e. the rate at which mass is transferred in a given direction per unit area normal to it ($kg\ m^{-2}\ s^{-1}$), is proportional to the concentration gradient in that direction. In other words,

$$j \; \alpha - \left(\frac{dC}{dx}\right) \; \text{or} \; J = -D\left(\frac{dC}{dx}\right) \tag{4.1}$$

where D, the constant of proportionality, is known as *diffusion coefficient* or *diffusivity* which has the unit: m^2s^{-1}. It may be noted that the concentration and flux can also be taken in molar units. But diffusivity will still have the same unit, i.e. m^2s^{-1}. In writing Eq. (4.1), we have assumed that the concentration is only a function of position. In any real situation, the concentration depends on position as well as time, and in such situations, the concentration gradient also varies. Under such general conditions, we resort to the application of Fick's second law, which, for unidirectional diffusion, can be represented as:

$$\frac{\partial C}{\partial t} = D\frac{\partial^2 C}{\partial x^2} \tag{4.2}$$

It may be noted that the diffusion coefficient has been assumed to be constant in the above equation. This cannot be taken for granted in food systems and Fick's second law, which is the basic equation for unidirectional diffusive mass transfer, may be written as:

$$\frac{\partial C}{\partial t} = \frac{\partial}{\partial x}\left(D\frac{\partial C}{\partial x}\right) \tag{4.3}$$

Given that the term within brackets in Eq. (4.3) is negative of the flux defined in Eq. (4.1), it can be said that the rate of change of concentration is equal to the negative gradient of flux. In order to determine concentration as a function of position and time, it is necessary to solve Eq. (4.3) which is clearly a partial differential equation, second order w.r.t x and first order w.r.t t. We therefore need two boundary conditions relating C with x and one relating C with t, i.e. three conditions in all. It may be noted that Eq. (4.3) is analogous to Eq. (3.15) for conductive heat transfer. If diffusion takes place through an infinite slab with constant surface concentration, the solution of Eq. (4.3) for *constant diffusion coefficient* is similar to Eq. (3.17) valid for conductive heat transfer; i.e.

$$\frac{C(x,t) - C_S}{C_i - C_S} = erf\left(\frac{x}{2\sqrt{Dt}}\right) \tag{4.4}$$

where $C(x,t)$ is the transient concentration at any position x; C_i is the initial concentration assumed to be uniform; and C_s is the constant surface concentration of the solute. Error function values are listed in Table 3.2.

Problem 4.1 A fresh apple slice, 5 mm thick with an initial uniform sucrose content of 150 kg m^{-3}, is exposed to water flowing at a very high speed around it. The concentration of sucrose on the surface of the slice may be assumed to be

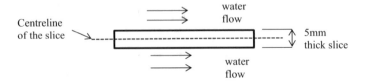

Fig. 4.1 Sucrose loss from an apple slice by diffusive mass transfer

negligibly low. If the diffusion coefficient of sucrose in the apple slice is 1.0×10^{-9} m^2 s^{-1}, what will be the sucrose concentration at a position 1 mm from the center of the slice after 10 min of exposure?

Solution Although the solution of this problem is based on the direct application of Eq. (4.4), it is necessary to appreciate some background principles. The sucrose diffuses from within the apple to its surface, because the surface concentration of sucrose is significantly lower. It is true that the flowing water will also offer resistance to mass transfer, but since it is flowing at a very high speed, its resistance can be considered to be negligible in relation to the diffusive resistance inside apple the slice. It is also necessary to consider that the sucrose will diffuse on either side of the centerline shown in Fig. 4.1. It is therefore necessary to note that the maximum diffusion path length for the sucrose is 2.5 mm. In relation to Eq. (4.4), $C_s = 0$, $C_i = 150$ kg m^{-3}, $x = 1$ mm $= 10^{-3}$ m, and $t = 10$ min $= 600$ s. Thus,

$\frac{C(x,t)-0}{150-0} = erf\left(\frac{10^{-3}}{2\sqrt{10^{-9} \times 600}}\right)$. The error function value, using Table 3.2, is approximately 0.64, whence $C(x,t) = 96.2$ kg m^{-3}.

The above problem illustrates the principle of diffusive extraction from fruit tissues to yield fruit juices such as apple juice. The same principle also applies to the extraction of sugar from sugar beet cossettes.

4.2.2 Steady State Diffusive Mass Transfer Through a Plane Membrane and Application to Diffusion Through Food Packaging Plastic Films

We will now consider some special cases of diffusion which are based on the solution of Eq. (4.2), i.e. where the diffusion coefficient, D, does not vary. The simplest case relates to *steady state* diffusion across a membrane of thickness L, which separates two well-stirred solutions having solute concentrations C_1 and C_2 ($C_1 > C_2$); see Fig. 4.2. Given that diffusion occurs under steady state conditions, the time derivative of concentration in Eq. (4.2) vanishes, and we have the following ordinary second order differential equation, with boundary conditions, $C = C_1$ at $x = 0$ and $C = C_2$ at $x = L$:

Fig. 4.2 Steady state diffusion across a membrane of thickness L, which separates two well-stirred solutions having solute concentrations C_1 and C_2 ($C_1 > C_2$). The solute concentration varies linearly along the membrane length

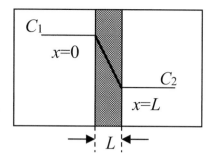

$$\frac{d^2C}{dx^2} = 0 \tag{4.5}$$

The solution of the above equation after incorporating the boundary conditions gives the following concentration profile inside the membrane:

$$C(x) = -\left(\frac{C_1 - C_2}{L}\right)x + C_1 \tag{4.6}$$

The above equation shows that the concentration varies linearly with x, and its gradient is given by:

$$\frac{dC}{dx} = -\frac{C_1 - C_2}{L} \tag{4.7}$$

Further, the mass transfer flux is given by:

$$J = -D\left(\frac{dC}{dx}\right)_{x=0} = D\left(\frac{C_1 - C_2}{L}\right) \text{ or } \left(\frac{D}{L}\right)\Delta C \tag{4.8}$$

where $(C_1 - C_2) = \Delta C$ the concentration difference.

In the above example, we have assumed that the two solutions are well-stirred. In other words, we have discounted the possibility of encountering concentration gradients in the two solutions which effectively suggests that the concentration at the two surfaces of the membrane correspond to the concentrations of the solutions with which they are in contact (i.e. $C = C_1$ at $x = 0$ and $C = C_2$ at $x = L$). Also implicit in this assumption is the fact that the membrane has no specific affinity for the solute. It is quite possible that the solute is able to partition between the solution and membrane, in which case, the concentrations at the surfaces are not C_1 and C_2, but mC_1 and mC_2, where m is the partition coefficient. In this situation, the boundary conditions for solving Eq. (4.5) get modified, but the solution, and expressions for concentration gradient and flux, are similar except that are C_1 and C_2 are replaced by mC_1 and mC_2.

A practical application of diffusive mass transfer across a thin film relates to the transfer of gases and vapours through food packaging films. For example, consider a hamburger patty that is wrapped in a plastic film of thickness L. The air in contact with the hamburger has oxygen, moisture and other components at concentrations which are different to their concentrations in the air outside the packaging. Thus, concentration gradients are set up across the plastic film through which diffusion can occur. Whether the diffusion occurs from the inner side to the outer side of the film, or *vice versa*, depends on the gradients. If we consider moisture, it is most likely that the moisture content inside the packaging would be greater than the moisture content outside, simply because the patty has a high moisture content which will yield a higher partial pressure of water vapor inside the packaging. If we consider oxygen, it is quite possible that food deterioration with time is predominantly oxidative and consumes oxygen present inside the package, which will result in a gradient where the oxygen concentration outside the packaging is greater than the concentration inside.

Let us consider moisture transfer from the inner side to the outer side of the packaging film – which can be schematically shown by Fig. 4.2 where the shaded membrane, now, represents the plastic film. The moisture flux will be given by Eq. (4.8) where C_1 and C_2 are the moisture contents on the two sides of the packaging surface, which will be thermodynamically related to the partial pressures of water vapor on either side of the film, p_1 and p_2. It is reasonable to assume that Henry's law of gas solubility applies; so $C_1 = p_1/H_w$ and $C_2 = p_2/H_w$, where H_w (Pa m^3 kg^{-1}) is the Henry's constant describing solubility of moisture in the plastic material from which the film has been made. The moisture transfer flux will therefore be given by Eq. (4.8) as follows:

$$J = D\left(\frac{p_1 - p_2}{LH_w}\right) = P\left(\frac{p_1 - p_2}{L}\right) = \left(\frac{p_1 - p_2}{L/P}\right) \tag{4.9}$$

where $P = D/H_w$ is known as the permeability of water vapour through the film (Pa^{-1} m^{-1} s^{-1}kg). Permeability is a term that is extensively used in food packaging design and optimization. Since J is the mass transfer flux in Eq. (4.9) and $(p_1 - p_2)$ is the driving force, L/P may be considered to be the resistance offered by the film to moisture diffusion. In other words, L/P is a measure of the barrier property of the film, in this case, with respect to moisture. Likewise, the same film will have different values of permeability, P_1, P_2 etc. for different components such as oxygen, carbon dioxide and other components, and correspondingly different barrier properties to the different components. If the packaging material is a composite of three materials, which is not uncommon, the net resistance to mass transfer for any given gaseous component will be $\Sigma\,(L/P)$.

Problem 4.2 Potato crisps (or chips) are packed in a well-sealed plastic bag flushed in Nitrogen at 1.05 atm pressure. The thickness of the plastic is 0.05 mm and its surface area is 0.1 m^2. If the permeability of the plastic film to Nitrogen is 1.72×10^{-21} Pa^{-1} m^{-1} s^{-1}kg, estimate the mass of nitrogen lost from the bag

over a period of 3 months or 90 days, assuming that the ambient total pressure is 1 atm and the volume fraction of nitrogen in ambient air is 0.8. Also note that 1 atm = 101,330 Pa.

Solution In the above problem, the permeability of nitrogen through the plastic film is $P = 1.72 \times 10^{-21}$ Pa^{-1} m^{-1} s^{-1}kg; $p_1 = 1.05$ atm, whereas $p_2 = 0.8$ atm; and finally, $L = 0.05$ mm $= 5 \times 10^{-5}$ m From Eq. (4.9), the mass flux of nitrogen is given by:

$$J = P\left(\frac{p_1 - p_2}{L}\right) = 1.72 \times 10^{-21} \left(\frac{(1.05-0.8)\times 101330}{0.05\times 10^{-3}}\right) = 8.71 \times 10^{-13} \text{ kg m}^{-2} \text{ s}^{-1}.$$

Given that the surface area of the bag through which diffusion can occur is $A = 0.1$ m^2, and the time period over which diffusion occurs is 3 months which can be converted into seconds, the total mass of nitrogen lost is $8.71 \times 10^{-13} \times 0.1 \times (90 \times 24 \times 3600) = 6.78 \times 10^{-7}$ kg, which roughly works out to 2.42×10^{-5} mols – which is a very small amount. So the pressure inside the bag would hardly vary.

Problem 4.3 Meat is packed under an oxygen free modified atmosphere in a virtually impermeable plastic punnet sealed with a thin PVC film on the top which allows oxygen from ambient air to diffuse into the punnet. What must be the thickness of the PVC film so that 3.2×10^{-6} kg oxygen enters the punnet per day. The surface area of the PVC film is 0.04m^2 and its permeability is 4.5×10^{-19} Pa^{-1} m^{-1} s^{-1}kg. Ambient air may be assumed to contain 20 mole % of oxygen.

The oxygen flux into the punnet from the air is $J = 3.2 \times 10^{-6}/(0.04 \times 24 \times 3600) = 9.26 \times 10^{-10}$ kgm^{-2} s^{-1}. The ambient partial pressure of oxygen $p_1 = 0.2$ atm while the partial pressure of oxygen inside the punnet $p_2 = 0$. Taking the total pressure atmospheric pressure to be 101,330 Pa, we have from Eq. (4.9): $9.26 \times 10^{-10} = 4.5 \times 10^{-19} \left(\frac{(0.2-0)\times 101330}{L}\right)$, whence $L = 9.8 \times 10^{-6}$ m.

It may be noted that the mass transfer flux J and the driving force for mass transfer have been assumed to remain constant during diffusion. This is reasonable because the amount of gas diffused is so low that it does not change the partial pressures on either side of the film.

4.2.3 Steady State Diffusive Mass Transfer in Spherical Systems

A second example of practical relevance relates to a spherical system, where an external solute diffuses into a sphere, or a solute from the sphere evaporates into the surroundings. Let us consider steady state diffusive evaporation from a sphere of diameter d, with surface concentration $C = C_1$. The first point to note is that diffusion is not unidirectional; it occurs uniformly in all spatial directions, with the solute radiating outwards. If, however, we choose to analyze the problem applying a spherical coordinate system, only one coordinate, i.e. the radial direction, is relevant. As a consequence, the equations describing diffusion will only have one independent variable – the radial position. If Q is the uniform outward rate of diffusion of solute from the sphere expressed in kg s^{-1}, the mass transfer flux at any radial position, r, measured from the centre of the sphere, will be $Q/(4\pi r^2)$. According to Fick's first law:

$$-D\frac{dC}{dr} = \frac{Q}{4\pi r^2} \tag{4.10}$$

Solving the above equation with the condition $C = C_1$ when $r = d/2$, the solute concentration at any radial position, C, is given by:

$$D(C_1 - C) = \frac{Q}{4\pi}\left(\frac{2}{d} - \frac{1}{r}\right) \tag{4.11}$$

Noting that the solute concentration $C \to 0$ when $r \to \infty$, the rate of mass transfer from the sphere, Q, can be shown from the above equation, to be:

$$Q = 2\pi d D C_1 \tag{4.12}$$

Further, given that the flux at any radial position is $Q/(4\pi r^2)$, the flux at the surface of the sphere, using Eq. (4.12), can be shown to be:

$$J_S = \frac{Q}{\pi d^2} = 2\left(\frac{D}{d}\right)C_1 \tag{4.13}$$

4.3 Interpretation of Diffusion Coefficient and Its Typical Values

Although diffusion coefficient or diffusivity was introduced in Sect. 4.2 as a constant of proportionality in the Fick's first law, its numerical value gives an indication of molecular mobility through any phase, and it is also known as the *molecular diffusion coefficient*. It follows that diffusion coefficients in gases would be the highest. Typically, D values are of the order of 10^{-5} m^2/s in gases, 10^{-9} m^2/s in liquids, and 10^{-13} m^2/s or lower in solids. It is also be noteworthy that molecular size inversely affects these values. For instance, the diffusion of oxygen through water at 20 °C is approximately 2×10^{-9} m^2/s whereas that of macromolecules such as proteins can be one or two magnitudes lower. Equations for estimating diffusion coefficient are available in literature, and these equations rely on knowing specific molecular parameters of the solute, physical properties of the solvent and parameters

Table 4.1 Atomic and structural diffusion volume increments

C	16.5	Cl	19.5
H	1.98	S	17.0
O	5.48	Aromatic ring	−20.2
N	5.69	Heterocyclic ring	−20.2

For each gas or vapour, its molecular formula is first written and the volume increment of each atom in the formula is taken from the values given in the Table and added up to give Σv. For example Σv for H_2O is $2(1.98) + 5.48 = 9.44$

reflecting interactions between the two. The equation commonly used to estimate diffusion coefficient in gases is given by Fuller et al. (1966):

$$D_{AB} = \frac{1.00 \times 10^{-7} T^{1.75} (1/M_A + 1/M_B)^{0.5}}{P\left[(\sum v_A)^{1/3} + (\sum v_B)^{1/3}\right]^2} \tag{4.14}$$

In the above equation, D_{AB} is the diffusion coefficient of gas A in B expressed in m^2/s; M_A and M_B are the molecular weights of the A and B respectively; T is the absolute temperature; P is the absolute pressure in atmospheres; and Σv_A and Σv_B are known as the atomic diffusion volumes, which, if not known for a gas, are estimated by adding the contributions of each atom in its chemical structure. The individual atomic contributions for common atoms are listed in Table 4.1.

Diffusion of solutes in liquids is very important in many food processes such as drying where, in addition to water, aromatic flavour and other components are lost by this mechanism. The equations for predicting diffusion coefficient for solutes in liquids are essentially semi-empirical. For relatively small solvent molecules, and solute molecules of molecular weight 1000 or greater, the equation used is known as the modified Stokes-Einstein equation:

$$D_{AB} = \frac{9.96 \times 10^{-16} T}{\mu V_A^{1/3}} \tag{4.15}$$

In the above equation, T is the absolute temperature (K); μ is the solution viscosity in Pas; V_A is the solute molar volume at its normal boiling point expressed in $m^3/kmol$; and D_{AB} is expressed in m^2/s. For dilute aqueous solutions, the following equation due to Polson (1950) may also be used:

$$D_{AB} = \frac{9.40 \times 10^{-15} T}{\mu M_A^{1/3}} \tag{4.16}$$

where M_A is the molecular weight of the solute (>1000). When a small solute is diffusing in dilute aqueous solutions, the most common equation used is known as the Wilke-Chang equation (see Treybal (1980)):

$$D_{AB} = 1.173 \times 10^{-16}(\varphi M_B)^{1/2} \frac{T}{\mu_B V_A^{0.6}} \qquad (4.17)$$

In the above equation ϕ is known as the association parameter of the solvent. For water, $\phi = 2.6$, and for ethanol $\phi = 1.9$; values for some other solvents are reported in the reference stated above (Treybal 1980). It is interesting to note from Eq. (4.17) that D_{AB} is directly proportional to the absolute temperature T and inversely proportional to the liquid viscosity μ_B; i.e. $(D_{AB}\,\mu/T)$ is constant for a given gas A diffusing in a liquid B.

4.4   Concept of Mass Transfer Coefficient

The approach taken so far is based on the assumption that mass transfer flux is proportional to the concentration gradient. An alternative, and somewhat approximate approach, is to assume that the flux is proportional to *concentration difference*, ΔC, instead of the concentration gradient. In other words,

$$J \alpha \Delta C \quad \text{or} \quad J = k_m \Delta C \qquad (4.18)$$

The constant k_m is known as the *mass transfer coefficient*. If J is expressed in $kgm^{-2}\,s^{-1}$ and C is expressed in kgm^{-3}, the unit of k_m will be ms^{-1}. It may be noted that the same approach was adopted earlier in Chap. 3, Sect. 3.2, to account for heat transfer flux. The overall rate of mass transfer (R) can be obtained by multiplying both sides of Eq. (4.18) by the interfacial area A (m^2). In other words,

$$R = JA = k_m A \Delta C \qquad (4.19)$$

In practice, the specific rate of mass transfer (r) is used quite extensively. This term is defined as the rate of mass transfer per unit volume of the system, and it can be obtained by dividing both sides of Eq. (4.19) by the system volume (V) expressed in m^3:

$$r = \frac{R}{V} = k_m \left(\frac{A}{V}\right) \Delta C = k_m a \Delta C \qquad (4.20)$$

The term $a = A/V$ is known as the specific area for mass transfer and it has the unit: m^{-1}. It is important to note that the exact interfacial area for mass transfer between two phases, A or a, is rarely known, and cannot be determined accurately. On the other hand, the rate of mass transfer and concentration driving forces can be measured accurately. Thus, even though k_m and a cannot be individually determined, their product $k_m a$ can be accurately measured. This product is known as the *overall volumetric mass transfer coefficient* which has the unit s^{-1}.

It is also worth noting that diffusion coefficient (D) and mass transfer coefficient (k_m) are two parameters used to characterise the same quantity, i.e. mass transfer flux. It is therefore obvious that the two are related. Comparing Eq. (4.20) with Eq. (4.8), which represents flux across a thin membrane, it is evident that:

$$k_m = \frac{D}{L} \quad \text{or} \quad \frac{k_m L}{D} = 1 \tag{4.21}$$

The term ($k_m L/D$) is dimensionless, and it is known as the Sherwood number, denoted as Sh. Thus, Sh = 1 for diffusion across a plane membrane, and $k_m \propto D$. Likewise, we can also compare Eq. (4.20) with the expression for flux for solute evaporation from a sphere, i.e. Eq. (4.12), to yield:

$$Sh = \frac{k_m d}{D} = 2 \tag{4.22}$$

It appears from the above discussion that the Sherwood number for molecular diffusion depends on the shape or geometry of the interface. This does not necessarily have to be the case: whether Sh = 1 or 2, or for that matter any other value, depends on the definition of the length scale. For instance, if the length scale of the sphere in Eq. (4.22) is defined in terms of radius, Sherwood number would be equal to 1. It is therefore important to note the definition of length scale, before using specific equations for calculating mass transfer coefficient given the molecular diffusion coefficient.

Problem 4.4 A microorganism, 1 μm in diameter, is swimming in a glucose solution having a concentration of 1 kg m^{-3} (Fig. 4.3). Estimate the maximum rate at which the microorganism will be able to consume glucose, assuming that glucose diffusivity in the solution is 6.7 × 10^{-10} m^2 s^{-1}.

Solution The rate of transfer of glucose, $R = k_m A C$. The area for mass transfer $A = \pi d^2 = \pi (1 \times 10^{-6})^2 = 3.14 \times 10^{-12}$ m^2 where $d = 1$ μm $= 1 \times 10^{-6}$ m. For this situation, the mass transfer coefficient is related to the diffusivity of glucose by: Sh $= \frac{k_m d}{D} = 2$. Taking diffusivity $D = 6.7 \times 10^{-10}$ m^2 s^{-1}, $k_m = 2 \times D/$

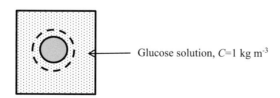

Glucose solution, C=1 kg m^{-3}

Fig. 4.3 An exaggerated representation of a microorganism, diameter $d = 1$ μm, in glucose solution. Diffusive mass transfer will occur from the bulk of the solution towards the interface of the microorganism through a hypothetical film around the microorganism shown by the broken line. The bulk concentration of glucose is C whereas the concentration at the microorganism surface may be assumed to be zero, since it consumes the glucose as soon as it comes into contact

$d = 1.34 \times 10^{-3}$ m s^{-1}. The rate of glucose transfer or glucose consumption $= (1.34 \times 10^{-3})(3.14 \times 10^{-12})(1) = 4.21 \times 10^{-15}$ kg s^{-1}.

4.5   Interfacial Mass Transfer

Practical applications of mass transfer theory involve movement of components across an interface. The method used to model mass transfer rates can be best illustrated by considering a process where a membrane separates two solutions and a solute diffuses across it. Figure 4.4a shows the solute distributed between two phases that are separated by the membrane. It is assumed that there are no concentration gradients in the *bulk* of the two phases, and the solute concentrations are uniform at C_1 and C_2 with $C_1 > C_2$. In the immediate vicinity of the membrane, however, the concentration on either side can be different: to the left of the interface, the solute concentration is lower than C_1, say C_{1i}; and to the right of the interface, the concentration is C_{2i} which is greater than C_2. The reason for these differences will be considered later. For the moment, it would suffice to note that the driving force for solute transfer to the left of the membrane is $(C_1 - C_{1i})$; and driving forces for transfer through the membrane and beyond are $(C_{1i} - C_{2i})$ and $(C_{2i} - C_2)$, respectively. The driving forces for mass transfer described here are very similar to the temperature driving forces for heat transfer described in Chap. 3, Sect. 3.1. If Eq. (4.20) is applied to describe the rate of mass transfer through each "resistance", and a pseudo steady state is assumed (i.e. there is no accumulation of solute in the membrane, and the extent of mass transfer is too low to affect solute concentrations on either side of the membrane), we have:

$$r = k_1 a_1 (C_1 - C_{1i}) = k_i a_i (C_{1i} - C_{2i}) = k_2 a_2 (C_{2i} - C_2) \qquad (4.23)$$

Further, if the rate of mass transfer is described in terms of an overall driving force (i.e. $(C_1 - C_2)$) and an overall mass transfer coefficient, $K_1 a_1$, as:

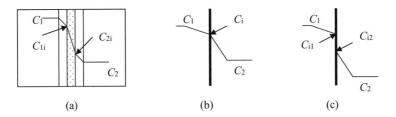

(a)   (b)   (c)

Fig. 4.4 (**a**) Interfacial mass transfer across a membrane of thickness L, which separates two well-stirred solutions having solute concentrations C_1 and C_2 ($C_1 > C_2$). The solute concentration varies linearly across the membrane. (**b**) Solute concentrations when the membrane shown in (**a**) is very thin. (**c**) Concentration profile on either side of a very thin membrane when concentrations in each phase is different

$$r = K_1 a_1 (C_1 - C_2) \, , \tag{4.24}$$

we can show, similar to Eq. (3.22), that:

$$\frac{1}{K_1 a_1} = \frac{1}{k_1 a_1} + \frac{1}{k_i a_i} + \frac{1}{k_2 a_2} \tag{4.25}$$

It may be noted that for mass transfer, a_1, a_2, and a_i are very nearly equal. Thus,

$$\frac{1}{K_1} = \frac{1}{k_1} + \frac{1}{k_i} + \frac{1}{k_2} \tag{4.26}$$

If the reciprocal of mass transfer coefficient represents the resistance to mass transfer, the overall resistance is the sum of individual resistances. In other words, the three resistances act in series. This conclusion is very similar to the one drawn for heat transfer in Chap. 3, Sect. 3.2).

If the membrane was very thin (Fig. 4.4b), its resistance would be negligible (i.e. $1/k_i \rightarrow 0$); and $C_{1i} = C_{2i}$. In other words, there is no driving force across the membrane, where a single uniform concentration, say, C_i exists.

It may be noted that the concentrations on either side of a thin membrane or "film" do not necessarily have to be the same (see Fig. 4.4c). This type of situation arises, for instance, when a protein transfers between two insoluble liquids (eg aqueous and organic liquids), where the transferring component partitions between the two phases. As evident in Fig. 4.4c, C_{i1} and C_{i2} are different but they are thermodynamically related, normally, by the equation:

$$C_{i2} = C_{i1}/m \tag{4.27}$$

where m is the distribution or partition coefficient. The driving force for mass transfer in phase 2 is: $(C_{i2} - C_2) = (C_{i1}/m - C_2)$. The driving force in phase 1 is simply $(C_1 - C_{i1})$. If these driving forces replace the ones in Eq. (4.23) and the term containing $k_i a_i$ is omitted, the following equation, similar to Eq. (4.26), relates the individual phase mass transfer coefficients to the overall mass transfer coefficient:

$$\frac{1}{K_1} = \frac{1}{k_1} + \frac{m}{k_2} \tag{4.28}$$

4.6 Convective Mass Transfer

We have so far considered diffusive mass transfer which has been induced by concentration differences between two positions. This mechanism will prevail if there was no bulk movement around the positions. This is very unlikely in practice, since materials are normally under constant convection, which may be induced by density differences (*natural convection*) or by mechanical agitation (*forced convection*). Mass transfer caused by convection tends to dominate over molecular diffusion. We shall now consider mass transfer across an interface formed by two phases that are in a state of convection. Although there are several theories describing convective transfer, we will consider the three important ones: the *film theory* (also known as *two film theory*), Higbie's *penetration theory* and Danckwerts' *surface renewal theory*.

4.6.1   Film Theory

Consider mass transfer occurring between two phases 1 and 2, as shown in Fig. 4.5. This theory postulates the presence of a film on either side of the interface where the flow is fully developed laminar, and convection ceases to exist in the direction of mass transfer. The thickness of the film depends on the intensity of convection in the bulk: the greater the intensity, the closer will convective currents get to the interface, and the film will be thinner. The entire resistance to mass transfer in a given phase is assumed to be concentrated in this film. We have already shown in Eq. (4.21) that the Sherwood number for molecular diffusion across a film is unity when the length scale is expressed in terms of the film thickness, and the same equation will therefore relate mass transfer coefficient with diffusion coefficient. Thus, the individual phase mass transfer coefficients in Fig. 4.5 are given by:

$$k_1 = \frac{D_1}{\delta_1} \quad \text{and} \quad k_2 = \frac{D_2}{\delta_2} \tag{4.29}$$

In other words, the mass transfer coefficient is proportional to diffusion coefficient. Although the film theory is not accurate, it is very simple and gives a reasonable description of some complex mass transfer problems.

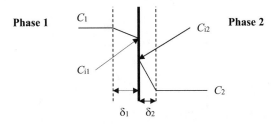

Fig. 4.5 Schematic illustration of the two-film theory with mass transfer occurring from phase 1 into phase 2. The film thickness is denoted by δ, and C represents the concentration of the transferring species. The subscripts refer to the respective bulk phases with i representing the interface. The figure shows $\delta_1 > \delta_2$, which suggests that the resistance offered by Phase 1 to mass transfer is greater than Phase 2

4.6.2   Higbie's Penetration Theory

The film theory assumes the film adjacent to the interface to be in a state of fully developed laminar flow. Higbie (1935) pointed out that contacts between phases often occurred repeatedly, over time scales that were too short for fully developed laminar flow to be established adjacent to interfaces. He also showed that the time averaged mass transfer coefficient depended upon the contact time, t, exposed to penetration:

$$k_1 = 2\sqrt{\frac{D_1}{\pi t}} \qquad (4.30)$$

Thus, the contact time, t, is the model defining parameter of penetration theory, just as δ was for the film theory. Furthermore, just like δ, the contact time cannot be easily measured. The penetration theory, unlike the film theory, concludes that $k \propto \sqrt{D}$, i.e. the mass transfer coefficient is proportional to the square root of diffusion coefficient.

4.6.3   Danckwerts' Surface Renewal Theory

Danckwerts (1951) extended penetration theory after allowing for turbulent eddies to bring masses of fresh liquid continually to the interface, where they were exposed for a finite length of time and renewed. Danckwerts assumed that every element of liquid had the same chance of being replaced, regardless of its age, and showed that:

$$k_1 = \sqrt{D_1 s} \qquad (4.31)$$

where s is the fractional rate of surface renewal having units of reciprocal time (s^{-1}). Once again, similar to δ and t in the above models, s cannot be easily measured. It may be noted that both penetration and surface renewal theories show that $k \propto \sqrt{D}$.

Problem 4.5 If a spherical air bubble of diameter 2 mm rises vertically through water at a velocity of 0.2 m s^{-1}, and its surface is renewed each time it traverses a distance equal to its diameter, estimate the mass transfer coefficient using Danckwerts' model if the diffusion coefficient for oxygen in water is 1.2 × 10^{-9} m^2 s^{-1}. What is the maximum possible rate of oxygen transfer from this bubble in kg s^{-1} if oxygen solubility in water is 10 mg per litre.

Note that the bubble diameter $d = 2$ mm $= 2 \times 10^{-3}$ m. Since the bubble surface is renewed each time the bubble rises a distance equal to its own diameter, the rate of surface renewal, $s = \frac{\text{bubble rise velocity}}{\text{bubble diameter}} = \frac{0.2}{2 \times 10^{-3}} = 100$ s^{-1}. The mass transfer coefficient, $k_1 = \sqrt{D_1 s} = \sqrt{(1.2 \times 10^{-9})(100)} = 3.46 \times 10^{-4}$ m s^{-1} . The rate of oxygen transfer $= k_1 A C^*$, where $A = \pi d^2$ is the bubble surface area across which mass transfer occurs, and $C^* = 10$ mg/l $= 10^{-2}$ kg m^{-3} the given solubility of oxygen in water. Note that the rate of transfer will be maximum when the bulk concentration in water is zero. Thus, the maximum rate of oxygen transfer $= (3.464 \times 10^{-4})\,(\pi\,(2 \times 10^{-3})^2) \times 10^{-2} = 4.35 \times 10^{-11}$ kg/s

Experimental investigations have shown that the exponent on D is neither 1 as suggested by the film theory, nor 0.5 as suggested by the penetration and surface renewal theories. A number of empirical equations relating k with D have been reported. Most equations for convective mass transfer follow what is commonly known as the Sherwood approach, and they relate k and D through the following dimensionless numbers in the case of *forced convection*: Sherwood number which represents the ratio of diffusive to convective resistance (Sh $= kD/d$ where d is the length scale), Reynolds number (Re $= du\rho/\mu$) already explained in Chap. 1, and Schmidt number (Sc $= \mu/\rho D$) which represents the ratio of momentum to mass diffusivity. For instance, the mass transfer coefficient for evaporation from a sphere of diameter d, when forced convection prevails around it is given by:

$$\text{Sh} = 2 + 0.552\ \text{Re}^{0.5}\text{Sc}^{0.33} \tag{4.32}$$

It may be noted that convection does not replace molecular diffusion but is superimposed upon it. Hence the first term of Eq. (4.32) is 2, which follows from Eq. (4.22). Such equations relating mass transfer coefficient with molecular diffusion coefficient are not only dependent on fluid properties, but also the geometry of the system. For instance, the equation for evaporation from flat plate geometry or cylindrical geometry will be similar to Eq. (4.32), except that the constants will be different. Further, the constants are also dependent on the values of Sc and Re. For instance, the constants in the second term of Eq. (4.32) may be used when $2 < \text{Re} < 800$ and $0.6 < \text{Sc} < 2.7$. Another correlation for forced convection mass transfer around spheres, which seems valid over very wide ranges of Re and Sc ($3.5 < \text{Re} < 7.6 \times 10^4$, and $0.71 < \text{Sc} < 380$), is:

$$\text{Sh} = 2 + \left(0.4\ \text{Re}^{0.5} + 0.06\,\text{Re}^{0.067}\right)\text{Sc}^{0.4} \tag{4.33}$$

If, instead of forced convection, *natural convection* prevailed around the evaporating sphere, the Reynolds number will not appear in the equation, instead, the Grashoff number will be introduced (Gr $= g\Delta\rho d^3/\mu^2$) where $\Delta\rho$ is the density difference between the sphere and the surrounding medium. Thus, for natural convection:

$$\text{Sh} = 2 + 0.43(\text{Gr}\ \text{Sc})^{0.25} \tag{4.34}$$

It may be noted that Eq. (4.34) is normally used when $1 < (\text{GrSc}) < 10^5$, and Sc takes values around 1. Acevedo et al. (2007) suggested that the constant 0.43 in Eq. (4.34) must be replaced by 1.5 to estimate mass transfer from fruits stored in packages. The similarity of equations used for estimating mass transfer coefficients with those for heat transfer coefficients (Table 3.3) may be noted. The Sherwood number for mass transfer is analogous to Nusselt number for heat transfer, and Schmidt number for mass transfer is analogous to Prandtl number for heat transfer. Momentum, heat and mass transfer are thus analogous, and their analogy can be exploited to gain insights into transport phenomena.

Problem 4.6 A raw (uncooked) meatball, 3 cm in diameter, is placed in an oven, where water evaporates by natural convection. If the air temperature around the meat ball is 60 °C, and the vapour pressure of water at this temperature is 20 kPa, estimate the rate at which water evaporates. If the meat ball is placed in a fan oven where the air velocity around the meat ball is 5 m s^{-1} and the temperature around it is 60 °C, what will be the rate of water evaporation? At 60 °C, the density of air is equal to 1.06 kg m^{-3}, and its viscosity is 2 × 10^{-5} Pas. The diffusion coefficient of water vapor at 60 °C may be assumed to be 2.8 × 10^{-5} m^2 s^{-1}.

Solution For natural convection, the mass transfer coefficient may be estimated by using Eq. (4.34) to evaluate Sh. Gr and Sc can be evaluated assuming $g = 9.8$ m s^{-2}, $\Delta\rho$ = the density difference between the meat and surrounding air 1000 kg m^{-3} (approx.) diameter $d = 0.03$ m, $\mu_{air} = 2 \times 10^{-5}$ Pas, $\rho_{air} = 1.06$ kg m^{-3} and diffusivity of water vapoer in air $D = 2.8 \times 10^{-5}$ m^2 s^{-1}. Substituting these values, Gr = 6.61×10^8 and Sc = 6.74; and from Eq. (4.33), Sh = 113.1. It is worth noting that the second term of Eq. (4.34) is significantly greater than 2 and neglecting to add the first term will result in no significant error! It may also be noted that the range of the value of GrSc = 4.45×10^9 which is outside of the range of applicability of Eq. (4.34); even Sc value of 6.74 is outside the range of applicability. Nevertheless, this equation is being used primarily to illustrate the method to evaluate mass transfer coefficient. Thus $k_1 d/D = 113.1$, whence the natural convective mass transfer coefficient $k_1 = 0.11$ m s^{-1}. Following the method mentioned in Problem 4.5, the rate of mass transfer = $k_1 A(C^* - 0)$. Here A is the mass transfer area of the sphere = $\pi d^2 = 2.8 \times 10^{-3}$ m^2. The driving force for mass transfer is the difference between the concentration of water vapor in the air at the surface of the meat ball, C^*, and that in the bulk air which can be assumed to be zero. The concentration of water vapour in the air at the surface of the meatball can be estimated by assuming that the air is saturated with water vapor, i.e. the partial pressure of water is equal to the vapour pressure. If p^* is the water vapour pressure (=20 kPa or 20×10^3 Pa), assuming ideal gas behavior, $C^* = p^*/RT$. With R, the ideal gas constant, taking the value 8.314 J gmol^{-1} K^{-1} and $T = 60 + 273 = 333$ K, $C^* = 7.22$ g mol of water vapour m$^{-3} = 0.13$ kg m^{-3}. Thus, the rate of mass transfer = $0.11 \times (2.8 \times 10^{-3})$ $(0.13) = 4.0 \times 10^{-5}$ kg s^{-1}.

An alternative correlation, which seems valid for higher values of the product (GrSc) such as those encountered in this problem, has been proposed by El Sherify and Hussein (1984): Sh = 0.15 (GrSc)$^{0.33}$. If we use this correlation, Sh = 0.15 $(6.61 \times 10^8 \times 6.74)^{0.33} = 229.2$ which is almost double the value predicted by Eq. (4.34). It is therefore critical to exercise caution while using such empirical correlations and it is necessary to check the applicability of any correlation for the values of the dimensionless numbers encountered.

When the meatball in the above problem is placed in the fan oven, forced convection dominates mass transfer. The Reynolds number is given by $Re = du\rho/\mu = (0.03)\,(5)\,(1.06)/2 \times 10^{-5} = 7950$. Substituting this value of Re and $Sc = 6.74$ in Eq. (4.33), $Sh = 131.24$. The mass transfer coefficient and mass transfer rates can now be estimated as before.

4.7   Oxygen Transfer Rates in an Aerobic Bioreactor

The solubility of oxygen in water under ambient conditions is a mere 10 mg per litre (the precise value depends on the temperature, pressure and water quality). This can pose severe oxygen shortages in aqueous systems which support respiring life in the form of microorganisms and aquatic creatures. These include natural water streams (rivers, lakes etc) and more intensive systems like fermenters, aquaculture and sewage treatment devices. The level of oxygen in such systems can only be sustained by sparging air or oxygen bubbles which transfer oxygen across their interfaces.

The rate of oxygen transfer from bubbles can be characterized by considering the schematic diagram illustrated in Fig. 4.6. Oxygen follows the concentration gradient and, first of all, transfers from the gas phase to the interface; and subsequently, it dissolves in the bulk liquid. The gaseous contents of the bubble are in convection or constant motion, and they can be assumed to be well-mixed throughout the bubble, except, perhaps, very close to the interface where a thin film can be assumed to exist. Since component concentrations in gaseous mixtures are normally given by partial pressures (Chap. 1, Sect. 1.2.2), the bulk and interfacial concentrations of oxygen can be can be represented by p_g and p_i; and the rate of oxygen transfer can be written as:

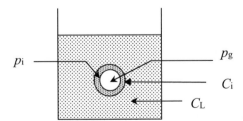

Fig. 4.6 Schematic illustration of oxygen transfer from an air bubble. p_g is the partial pressure of oxygen in the bubble; p_i is the partial pressure of oxygen at the interfacial film; C_i is the oxygen concentration at the interface, assumed to be in equilibrium with p_i (i.e. $C_i = p_i/H^*$ where H^* is the Henry's constant); and C_L is the bulk concentration of dissolved oxygen

$$r_g = k_g a (p_g - p_i) \qquad (4.35)$$

Note that if r_g is expressed in kg m^{-3} s^{-1}, a (the specific interfacial area) in m^{-1}, k_g the gas phase mass transfer coefficient, will have the unit kg m^{-2} s^{-1} Pa^{-1}. Likewise, the oxygen concentrations in the bulk liquid and at the interface can be represented by C_L and C_i; and the rate of oxygen transfer is given by:

$$r_L = k_L a (C_i - C_L) \qquad (4.36)$$

where k_L the liquid phase mass transfer coefficient, has the unit m s^{-1}. If we further assume that the interfacial film is very thin, the schematic shown in Fig. 4.6 is similar to the one shown in Fig 4.4b, except that the interface in the latter is shown to be rectangular. It is reasonable to assume that the interfacial concentrations of the gas and liquid phases are in thermodynamic equilibrium, i.e. the liquid surface is saturated with oxygen or C_i and p_i are related by Henry's law:

$$C_i = p_i / H^* \qquad (4.37)$$

where H^* is the Henry's constant (Pa kg^{-1} m^3). Eq. (4.37) is similar to Eq. (4.27), and the overall mass transfer coefficient K_L for this situation can be written in a form similar to Eq. (4.28):

$$\frac{1}{K_L} = \frac{1}{k_L} + \frac{1}{H^* k_g} \qquad (4.38)$$

In practice, the gas phase resistance to mass transfer is much lower than the liquid phase resistance. As a consequence, $p_g = p_i$. Further, the gas phase transfer coefficient is significantly greater than the liquid phase transfer coefficient, and the second term of Eq. (4.38) can be neglected. Thus, the rate of oxygen transfer is given by:

$$r = k_L a (C^* - C_L) = k_L a (p_g / H^* - C_L) \qquad (4.39)$$

A number of empirical expressions, similar to Eq. (4.32), are available to estimate mass transfer coefficient (k_L) from bubbles, which is a case of forced convective mass transfer around a (nearly) spherical bubble. It is equally important to note that even though k_L can be estimated this way, a is more often not amenable to accurate determination. As mentioned earlier in Sect. 4.4, this problem can be surmounted by determining the product $k_L a$. Most experimental methods for the determination of $k_L a$ are based on measuring the transient values of dissolved oxygen concentrations, when air or oxygen is sparged continuously through a stirred liquid. If the liquid is assumed to be perfectly mixed, the rate of oxygen transfer from the gas phase per unit volume of the dispersion, r, is equal to dC_L/dt, and Eq. (4.39) can be written as:

$$\frac{dC_L}{dt} = k_L a(C^* - C_L) \tag{4.40}$$

which is an ordinary differential equation that can be solved with the help of the initial condition, $C_L = C_{L0}$ when $t = 0$, to give:

$$\ln\left(\frac{C^* - C_{L0}}{C^* - C_L}\right) = k_L a t \tag{4.41}$$

The term on the left hand side of Eq. (4.41) is plotted against time to yield $k_L a$ as the gradient of the best fit line.

Problem 4.7 **In an experimental investigation, air was sparged into a sterile medium. The dissolved oxygen concentration increased as shown below from an initial value $C_{L0} = 5 \times 10^{-4}$ kg m^{-3}:**

Time since air was switched on, t (s)	30	60	90	120
Oxygen concentration ($C_L \times 10^3$ kg m^{-3})	1.90	3.80	5.55	6.50

Estimate $k_L a$ if the solubility of oxygen in the medium $C^* = 10^{-2}$ kg m^{-3}.

Solution The above problem is a direct application of Eq. (4.41) and it can be easily solved using Excel by plotting $\ln\left(\frac{C^* - C_{L0}}{C^* - C_L}\right)$ against t and finding the gradient of the best fit line, which, when done, results in $k_L a = 8.1 \times 10^{-3}$ s^{-1}.

The values of $k_L a$ determined under a variety of operating conditions are correlated empirically for different bioreactor configurations of mass transfer equipment. For example, the $k_L a$ values for a typical impeller agitated aerobic fermenter is related to the power dissipation level per unit volume of the fermenter as well as the air flow rate expressed, either as the superficial gas velocity (ms^{-1}) in the fermenter or the volume of gas introduced per unit volume of the fermenter per minute (also known as vvm, denoted by Q). One of the most general correlations for $k_L a$ in *impeller agitated* gas-liquid reactors (a typical fermenter would fall in this category of reactors) has been proposed by Yawalkar et al. (2002):

$$k_L a = 0.0558\left(\frac{N}{N_{CD}}\right)^{1.464} Q T^{1.05} \tag{4.42}$$

which, in terms of gas velocity v_g (ms^{-1}), is roughly equivalent to:

$$k_L a = 3.35\left(\frac{N}{N_{CD}}\right)^{1.464} v_g \tag{4.43}$$

It is necessary to note that the above equations are *dimensional*, i.e. the constants are only valid when the various terms are taken in very specific units – which must be stated by authors proposing such equations. Equations (4.42) and (4.43) yield values of $k_L a$ in s^{-1} when N, the impeller speed is taken in revolutions per second or r.p.s.; Q, the vvm will have units of min^{-1}; T, the reactor diameter is in m; and v_g. the gas velocity is taken in m s^{-1}. The parameter N_{CD} in the above equations represents the *critical impeller speed* for gas dispersion. This is the minimum number of revolutions per second at which the impeller has turn in order to disperse the gas that is sparged into smaller bubbles. Below this critical rotational speed, the sparged gas would not be significantly affected by the impeller and would simply bypass its action. In order to estimate $k_L a$ using Eqs. (4.42) or (4.43), the value of N_{CD} must be known for a given system, which is also given by empirical equations for specific reactor configurations. For example, if a reactor is agitated by a *disk turbine* impeller in a reactor of standard configurations (see Yawalkar et al. 2002), N_{CD} is given by:

$$N_{CD} = \frac{4Q_G^{0.5}T^{2.5}}{D^2} \tag{4.44}$$

where Q_G is the volumetric gas flowrate (the product of v_g and the cross sectional area of the reactor $(\pi/4)T^2$, $m^3 s^{-1}$) and D is the impeller diameter (m). Note that normally the impeller diameter is either one-third or one-half of the tank diameter. Such dimensional empirical equations have been published for different types of reactors such as bubble columns, air-lift reactors etc. as well as different configurations in each type. It is also necessary to note that fermenters used in food and bioprocessing operations contain interfacially active agents, which may be naturally present, or added, say, to reduce or eliminate foaming. The presence of such agents significantly influences the gas-liquid dispersion characteristics, and therefore the values of N_{CD} and $k_L a$. It is therefore necessary to ensure that the right equations are sought out from the published literature while trying to estimate N_{CD} and $k_L a$.

References

Acevedo C, Sánchez E, Young ME (2007) Heat and mass transfer coefficients for natural convection in fruit packages. J Food Eng 80:655–661

Danckwerts PV (1951) Significance of liquid-film coefficients in gas absorption. Ind Eng Chem 43(6):1460–1467. https://doi.org/10.1021/ie50498a055

El Sherify TH, Hussein AK (1984) Natural convection mass transfer at spheres. J Appl Electrochem 14:91–95

Fuller EN, Schettler PD, Giddings JC (1966) New method for prediction of binary gas-phase diffusion coefficients. Ind Eng Chem 58:18–27. https://doi.org/10.1021/ie50677a007

Higbie R (1935) The rate of absorption of a pure gas into a still liquid during short periods of exposure. Trans Am Inst Chem Eng 31:365–389

Polson A (1950) Some aspects of diffusion in solution and a definition of a colloidal particle. J Phys Colloid Chem 54(5):649–652. https://doi.org/10.1021/j150479a007

Treybal RE (1980) Mass transfer operations, 3rd edn. McGraw-Hill, Singapore

Yawalkar AA, Heesink ABM, Versteeg GF, Pangarkar VG (2002) Gas–liquid mass transfer coefficient in stirred tank reactors. Can J Chem Eng 80:840–848

Further Reading

Chhabra RP, Shankar V (2017) Coulson and Richardson's chemical engineering. Volume 1B heat and mass transfer: fundamentals and applications, 7th edn. Elsevier, Oxford

Datta AK. Biological and bioenvironmental heat and mass transfer. Marcel Dekker Inc. ISBN 0-8247-0775-3

Chapter 5
Reaction Kinetics

Keshavan Niranjan

Aim This chapter deals with the models used to describe the rates of chemical and biochemical reactions encountered in food systems. The chapter covers homogeneous phase reactions as well as selected heterogeneous reactions relevant to food and bioprocessing. The principles of mass transfer covered in the previous chapter will be used to describe the rates at which reactant migrate between phases and the mass transfer rates will be combined with reaction kinetics in order to describe the overall rates in heterogeneous systems.

5.1 Introduction

Chemical and biochemical reactions inevitably occur when food is processed. In some cases, the reactions are desirable and may even be necessary. For example, the gelatinisation of starch during the cooking of rice, the browning of baked products, the softening of meat and generation of flavours during cooking. In such cases, the processing conditions must promote and facilitate the reactions. On the other hand, there are reactions such as protein denaturation, vitamin degradation, off-flavour formation which are undesirable. It is necessary to note that, both, desirable and undesirable reactions occurring in foods involve reactants which are naturally present in the foods. In many cases, the same reactants are responsible for desirable as well as undesirable reactions. Moreover, some of these reactions are catalysed by enzymes – which are also naturally present in the foods. It may be noted that the reactants present in foods do not simply react in one single step and stop. Most of these reactions are a network of several reactions occurring in series and parallel. It is

K. Niranjan
Department of Food and Nutritional Sciences, University of Reading, Reading, Berkshire, UK
e-mail: afsniran@reading.ac.uk

© Springer Nature Switzerland AG 2022
K. Niranjan, *Engineering Principles for Food Process and Product Realization*,
Food Engineering Series, https://doi.org/10.1007/978-3-031-07570-4_5

therefore critical for us to understand the nature of the reactions, identify the conditions which promote the desirable reactions but do not favour the undesirable reactions. Such complex reaction networks can only be analysed by modelling the rates of each reaction within a network and combining these individual rates to create an overall network model.

Browning of foods involves a variety of reactions. Caramelisation and Maillard reactions cause browning to occur. Caramelisation involves loss of water, followed by isomerisation and polymerisation of sugars. Maillard reaction, on the other hand, occurs as a result of the amino group from proteins reacting with reducing sugars when the food is heated. It may be noted that most foods naturally contain proteins and reducing sugars for such reactions to occur. Some of the Maillard reaction products may be responsible for favourable colour and flavour, while other Maillard reaction products may be undesirable. For instance, at relatively high temperatures, such as those encountered while toasting cereals, Maillard reaction can result in the formation of acrylamides which can be carcinogenic. While caramelisation and Maillard reaction are promoted by heat, there are other browning reactions which are catalysed by enzymes. For example, browning of fresh fruits and vegetables – which is undesirable – occurs when phenols present in these materials is converted to quinones in the presence of an enzyme called polyphenol oxidase. The process conditions such as concentrations of the various reactants, temperature, pH, the activity of water, and the activity of the enzyme can all influence the reactions which occur within the food. This chapter considers how chemical and biochemical reactions are modelled to describe the influence of all these parameters.

5.2   Kinetics of Chemical Reactions

The simplest chemical reaction can be represented as

A → Products

The rate at which this reaction occurs *at any given temperature and other reaction conditions* (such as pH) can be conveniently represented by the power law:

$$r = kC^n \tag{5.1}$$

The rate of the reaction, r, has the units of kg m^{-3} s^{-1}, i.e. it generally represents how fast the reaction occurs in a unit volume of the reaction mixture; and C represents the concentration of the reactant A expressed in kg m^{-3}. The model parameters k and n are obtained by fitting Eq. (5.1) to experimental data determined in a laboratory. k is known as the rate constant and n is known as the order of the reaction. Further, the order of the reaction has no units but the rate constant does, and its unit depends on n. It can easily be shown from Eq. (5.1) that the rate constant has the unit: kg^{1-n} m$^{-3(1-n)}$ s^{-1}. It may also be noted that n can take any value, integral or otherwise, but when it takes the value zero, the reaction is known as a zero order reaction, when it takes the value 1, the reaction becomes first order, and when it takes the value 2, the reaction becomes second order. It follows that the rate constant of a zero order reaction has the unit $kg\ m^{-3}\ s^{-1}$, that of a first order reaction has the unit s^{-1}, and so on.

It is also interesting to note that the rate and concentration units are based on the mass of the reactants. Chemists often use the molar mass instead of kg, and this is because the chemical nature of the reactant A is precisely known to them. In food and biological processes, it is very rare that the exact molecular formula (or molecular weight) of the reactants are known. Moreover, the reactant A may not even be a single species, but a group of species lumped together to facilitate modelling. In such cases, it helps to use mass based concentrations and rates. Further, the rate of the reaction, by definition, represents the rate at which the reactant is consumed or the products are formed – which are also related by stoichiometry. In other words,

$$r = -\frac{dC}{dt} \tag{5.2}$$

Combination of Eqs. (5.1) and (5.2) results in a first order ordinary differential equation which can be solved for all values of n except n = 1, using the initial condition, $C = C_o$ when $t = 0$, to yield the following equation which allows us to estimate the reactant concentration at any time t, if k and n are known.

$$\frac{C_0^{1-n} - C^{1-n}}{1-n} = kt \tag{5.3}$$

When n = 1, i.e. for a first order reaction:

$$\ln\left(\frac{C_0}{C}\right) = kt \text{ or } C = C_0 e^{-kt} \tag{5.4}$$

The term *conversion* is often used to represent the fraction (or %) of the reactant transformed during the reaction. If X is the fractional conversion, then $C_0 X$ ($kg\ m^{-3}$) has got converted and the amount of reactant remaining after time t is $C = C_0(1 - X)$. Thus, Eqs. (5.3), and (5.4) can be re-written in terms of X as follows:

$$\frac{C_0^{1-n}\left(1 - (1-X)^{1-n}\right)}{1-n} = kt \tag{5.5}$$

$$\ln\left(\frac{1}{1-X}\right) = kt \tag{5.6}$$

The time taken for the conversion to attain a value of 0.5 is known as the half-life ($t_{1/2}$). According to Eq. (5.6), the half–life of a first order reaction is given by:

$$t_{1/2} = \frac{\ln 2}{k} = \frac{0.693}{k} \tag{5.7}$$

5.2.1 Effect of Temperature on Reaction Kinetics: The Arrhenius Equation

It is necessary to note that the power law model (Eq. 5.2) and subsequent Eqs. (5.2, 5.3, 5.4, 5.5, and 5.6) are all valid when the reaction temperature is held constant. In other words, these equations represent *isothermal* kinetics. It is well known that reactions occur much faster at higher temperatures. The effect of temperature, T (K), on reaction rates is accounted for by assuming that the rate constant k varies with temperature according to Arrhenius equation:

$$k = A \exp\left(-\frac{E}{RT}\right) \quad \text{or} \quad \ln k = \ln A - \left(\frac{E}{R}\right)\frac{1}{T} \tag{5.8}$$

where A is commonly referred to as the *pre-exponential factor* or the *frequency factor*; E is the activation energy (J mol^{-1}) and R is the gas constant (8.33 J mol^{-1} K^{-1}). The term $E/(RT)$ is dimensionless, and it does not matter whether the molecular weight is known or unknown, since this term gets cancelled when E is divided by R. The activation energy reflects the sensitivity of the rate constant (and indeed the reaction itself) to temperature. Higher values of E imply that the reaction rate is more sensitive to changes in temperature. The activation energy is estimated by experimentally determining the reaction rate constant (k) at different temperatures (T), and finding the gradient of the best-fit line to the plot of ln$k$ against $1/T$ (see Eq. (5.8)).

Problem 5.1 Vitamin C (i.e. Ascorbic acid) degradation follows first order kinetics during the pasteurisation of pomegranate juice. The activation energy, and the rate constant for this degradation at 80 °C, were experimentally determined to be 82 kJ mol^{-1} and 5.8×10^{-5} s^{-1}, respectively. If the pomegranate juice is pasteurised at 90 °C for 2 min, what fraction of vitamin C will be degraded in the process? If the rate constant for the first order thermal degradation of polyphenol oxidase in the juice at 90 °C is 2 min^{-1}, what fraction of the enzyme is destroyed?

Solution From Eq. (5.8): $\frac{k_{90}}{k_{80}} = \exp\left[-\frac{E}{R}\left(\frac{1}{90+273} - \frac{1}{80+273}\right)\right]$. Given that $k_{80} = 5.8 \times 10^{-5}$ s^{-1}, $E = 82{,}000$ J mol^{-1} and R $= 8.33$ J mol^{-1}, $k_{90} = 1.25 \times 10^{-4}$ s^{-1}. The extent of vitamin C degradation is the value of X deduced from Eq. (5.6) by taking $k_{90} = 1.25 \times 10^{-4}$ s^{-1} and $t = 120$ s, which results in $X = 0.015$. Thus, only 1.5% of vitamin present in the juice is lost during pasteurisation.

The rate constant for the degradation of polyphenol oxidase is 2 min^{-1} $= 0.03$ s^{-1}. Using this value of k in Eq. (5.6) with $t = 120$ s, $X = 0.97$. In other words, 97% of this enzyme is destroyed during pasteurisation – which is one of the main purposes of the process. The destruction of the enzyme prevents the degradation of polyphenols in the fruit juice and maintains its antioxidant activity when consumed.

5.3 Multiple Reactions

The reaction considered so far in Sect. 5.2 has one single reactant which undergoes a single reaction to form the product. Food systems invariably involve multiple reactions, which may either occur in series or in parallel, or may constitute a react network consisting of series and parallel reactions.

5.3.1 Parallel Reactions

In the reaction scheme shown in Fig. 5.1, reactant A undergoes two separate first order reactions simultaneously. In a practical food process, one of these reactions may be desirable, while the other may not be. In terms of modelling the reaction rates, it is not sufficient to know the conversion of A (X) because this term gives no indication of how much A has been desirably converted. The term *selectivity* (S) is defined to indicate the progress of one reaction in relation to the other. At any instant, the selectivity of B with respect to C is the ratio of the rates at which B and C are formed.

$$S_{BC} = \frac{r_{A \to B}}{r_{A \to C}} = \frac{k_2 C_A}{k_1 C_A} = \frac{k_2}{k_1} \tag{5.9}$$

The final selectivity will simply be the ratio of the amount of A consumed to form B and that consumed to form C. It is interesting to note from Eq. (5.9) that the selectivity can be increased by increasing k_2 in relation to k_1, and this can be done by adjusting the temperature once the Arrhenius equation constants are known. The *yield* with respect to B represents the fraction of A reacted, which is converted to B. Based on Eq. (5.4), the amount of A converted to B in time t is: $C_{A0}\exp(-k_2 t)$. Likewise, the amount of A converted to C is $C_{A0}\exp(-k_1 t)$. The yield of the reaction leading to B is therefore given by:

$$Y_{A \to B} = \frac{C_{A0} e^{-k_2 t}}{C_{A0} e^{-k_1 t} + C_{A0} e^{-k_2 t}} = \frac{e^{-k_2 t}}{e^{-k_1 t} + e^{-k_2 t}} \tag{5.10}$$

Fig. 5.1 Reactant A undergoing two first order reactions in parallel with rate constants k_1 and k_2

5.3.2 Reactions in Series

The rates of the two reactions may be represented as: $r_{A \to B} = k_1 C_A$; and $r_{B \to C} = k_2 C_B$. Clearly B is formed by the first reaction and it is consumed in the second. So a mass balance equation for B can be written by considering the fact that the net rate of accumulation of B is equal to the rate of its formation minus the rate of its consumption (Fig. 5.2):

$$\frac{dC_B}{dt} = k_1 C_A - k_2 C_B = k_1 C_{A0} e^{-k_1 t} - k_2 C_B \qquad (5.11)$$

It may be noted that C_A in the above equation has been substituted by the expression given in Eq. (5.4). Assuming that there is no B present initially, i.e. $C_B = 0$ when $t = 0$, the differential Eq. (5.11) can be analytically solved to yield:

$$C_B = \frac{k_1 C_{A0}}{k_2 - k_1} \left(e^{-k_1 t} - e^{-k_2 t} \right) \qquad (5.12)$$

It is worth noting that the concentration of A progressively decreases with time, while the concentration of C progressively increases with time. But species B is formed in the reaction as well as consumed, so its concentration can either go up or down depending on the stage of the reaction. In fact the concentration of B increases initially, goes through a turning point and then decreases. These trends are illustrated in Fig. 5.3. The time when C_B attains a maximum value can be deduced by setting $dC_B/dt = 0$, where C_B is given by Eq. (5.12), and solving for t to give t_{opt}:

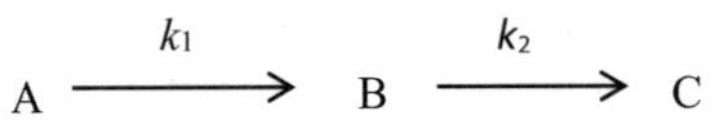

$$A \xrightarrow{\;k_1\;} B \xrightarrow{\;k_2\;} C$$

Fig. 5.2 Reactant A undergoes a first order reaction to form B which, in turn, undergoes another first order reaction to form C, the two reactions occurring in series with rate constants k_1 and k_2

Fig. 5.3 Concentration profiles of each species in a system where first order reactions $A \to B \to C$ occur in series, with k_1 and k_2 as the two rate constants, respectively

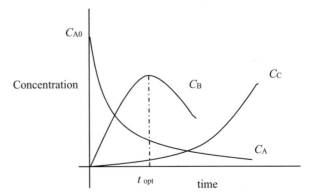

$$t_{opt} = \frac{1}{k_2 - k_1} \ln \frac{k_2}{k_1} \qquad (5.13)$$

The concentration of species C_C at any time is $(C_{A0} - C_A) - C_B$, where the term within brackets represents the total amount of B formed, and C_B is the amount of B present at time t given by Eq. (5.12):

$$C_C = \frac{C_{A0}}{k_2 - k_1} \left(k_2 \left(1 - e^{-k_1 t}\right) - k_1 \left(1 - e^{-k_2 t}\right) \right) \qquad (5.14)$$

Problem 5.2 Ascorbic acid (AA) in stored fruit juices degrades in the presence of oxygen to form dehydraoascorbic acid (DHA), which in turn degrades further to form 2,3-diketogulonic acid (DKG). Schematically, this reaction can be represented as: AA → DHA → DKG. At a given oxygen concentration, the two first order rate constants are $k_1 = 0.03 d^{-1}$ and $k_2 = 0.25 d^{-1}$, respectively. When does the concentration of DHA peak in the system and what is its maximum concentration if the initial AA concentration is 500 mg L^{-1}? What can you say about the rates of the two reactions simply by examining the rate constant values?

Solution The maximum concentration of DHA occurs when t is given by Eq. (5.13). With $k_1 = 0.03 d^{-1}$ and $k_2 = 0.25 d^{-1}$, $t = 9.64$ days. The maximum concentration of DHA can now be estimated from Eq. (5.12) with $k_1 = 0.03 d^{-1}$, $k_2 = 0.25 d^{-1}$, $t = 9.64$ days and $C_{A0} = 500$ mg L^{-1}, to give: $C_{DHA} = 44.94$ mg L^{-1}.

Comparing the rate constants $k_1 = 0.03 d^{-1}$ and $k_2 = 0.25 d^{-1}$, it is clear that the degradation of AA is significantly slower than the degradation of DHA. In a series of steps, it is the *slowest step* which controls the overall rate of the process. Thus, for all practical purposes, the degradation of AA indicates the rate of the overall process.

5.4 Kinetics of Enzyme Catalysed Reaction

Virtually every chemical reaction supporting life requires an enzyme in order to occur at a significant rate. Enzyme-catalysed reactions are millions of times faster than the corresponding uncatalysed reactions, and achieve this by *lowering the activation energy*. Each enzyme is also very specific, and operates on the reactant (commonly referred to as the *substrate*) over a very narrow range of reaction conditions, i.e. temperature and pH. Enzymes generally work by having an active site that is carefully designed by nature to bind a particular substrate. The interaction between the substrate and enzyme shift the substrate geometry closer to that of the transition state for the reaction. Once reaction has occurred, the products are released from the enzyme.

The kinetics of an enzyme catalysed reaction can be modelled by hypothesising the substrate S to form a complex ES with an enzyme E. The complex ES can convert to the product P and release E, or reverse its own formation process by returning the enzyme and the substrate. This hypothesis can be schematically represented as:

$$E + S \underset{k_1, k_{-1}}{\Leftrightarrow} ES \underset{k_2}{\rightarrow} E + P$$

where k_1 and k_{-1} represent the rates of the forward and backward reversible reaction forming the complex ES, and k_2 is the rate constant for the conversion of the enzyme into product. A mass balance for the species ES will consider the fact that its net rate of accumulation is zero in the system. In other words, its rate of formation from E and S must be equal to its rate of disappearance by the reverse reaction combined with its rate of conversion into product. Thus,

$$\frac{d[ES]}{dt} = k_1[E][S] - (k_{-1} + k_2)[ES] = 0 \quad \text{or} \quad [ES] = \frac{k_1[E][S]}{(k_{-1} + k_2)} \tag{5.15}$$

At the same time, a mass balance on the enzyme indicates that the amount of enzyme remaining must be equal to the amount of enzyme initially taken minus the amount of complex ES formed, or

$$[E] = [E]_0 - [ES] \tag{5.16}$$

Substituting the above expression for E in Eq. (5.15), it can be shown that:

$$[ES] = \frac{k_1[E]_0[S]}{k_{-1} + k_2 + k_1[S]} \tag{5.17}$$

The rate of the reaction, v is essentially the rate at which the product P is formed, i.e.

$$v = k_2[ES] = \frac{k_1 k_2 [E]_0 [S]}{k_{-1} + k_2 + k_1[S]} = \frac{k_2 [E]_0 [S]}{\left(\frac{k_{-1} + k_2}{k_1}\right) + [S]} \tag{5.18}$$

In Eq. (5.18), the term $k_2[E]_0$ is the maximum possible value of the rate of the reaction, say v_{max}, for a given initial enzyme concentration. Further, if $[(k_{-1} + k_2)/k_1] = k_M$, the rate of the enzyme catalysed reaction can be written as:

$$v = \frac{v_{max}[S]}{k_M + [S]} \tag{5.19}$$

Equation (5.19) is known as the *Michaelis-Menten equation* and is extensively used to correlate rates of enzyme catalysed reactions in food and bioprocessing. The term

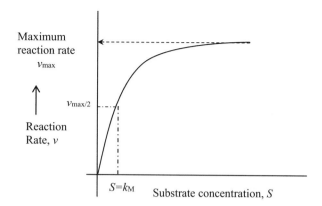

Fig. 5.4 Michaelis Menten plot of the rate of an enzyme catalysed reaction against substrate concentration. The Michaelis constant k_M is the substrate concentration at which the reaction rate is $v_{max}/2$. Further, the reaction is first order when S is low, whereas it tends to be zero order when S is high

k_M is also known as the Michaelis constant. Two extreme situations can be envisaged in Eq. (5.19). If $[S] \gg k_M$ or the substrate concentration is very high, $(k_M + [S]) \approx [S]$, which implies that the rate of the enzyme catalysed reaction is *independent of the substrate concentration*, i.e. it behaves like a zero order reaction. If on the other hand, the substrate concentration is low, say, $[S] \ll k_M$ then $(k_M + [S]) \approx k_M$. In other words, the reaction follows first order kinetics. The two model parameters v_{max} and k_M can easily be interpreted on the basis of the Michaelis Menten model (Eq. 5.19). While the former represents the maximum possible reaction rate under a given set of conditions, the latter represents the value of S when the reaction rate is $v_{max}/2$, which is clearly shown in Fig. 5.4. Just to simplify, $[S]$ has been replaced by S in the discussions below.

Equation 5.19 represents the rate of the enzyme catalysed reaction, so it can be used to develop expressions for the substrate concentration or substrate conversion at any time by noting that $v = -\frac{dS}{dt}$ and solving the differential equation with the initial condition: $S = S_0$ at t = 0. The resulting expression for S as a function of time is:

$$k_M \ln \left(\frac{S_0}{S}\right) + (S_0 - S) = v_{max} t \tag{5.20}$$

Alternatively, Eq. (5.20) can also be represented in terms of the fractional substrate conversion X (similar to Eqs. (5.20) or (5.21), noting that $S = S_0(1 - X)$, to yield:

$$k_M \ln \left(\frac{1}{1 - X}\right) + S_0 X = v_{max} t \quad \text{or} \quad S_0 X - k_M \ln (1 - X) = v_{max} t \tag{5.21}$$

It is necessary to note that the Michaelis Menten model is the simplest model representing enzyme catalysed reaction kinetics. Although its simplicity is its attraction, there are many situations where it is not applicable. For instance, the model assumes that the enzyme concentration does not change during the reaction. This need not be the case. There are many reactions involving more than one substrate, when this model cannot be directly applied. Some reactions are inhibited by the product formed. Michaelis Menten model does not account for product inhibition. Alternative models are available to deal with such situations.

Problem 5.3 The hydrolysis of Methyl Hydrocinnamate is catalysed by the enzyme chromotrypsin. If the k_M and v_{max} for this reaction are 0.003 mol dm^{-3} and 3.0×10^{-7} mol dm^{-3} s^{-1}, respectively, calculate the time required to achieve 75% conversion of Methyl Hydrocinnamate in a solution of strength 0.015 mol dm^{-3}.

Solution This problem involves direct application of either form of Eq. (5.21) where $X = 0.75$, $k_M = 0.003$ mol dm^{-3}; $v_{max} = 3.0 \times 10^{-7}$ mol dm^{-3} s^{-1}; and $S_0 = 0.015$ mol dm^{-3}. Substituting these values in Eq. (5.21), we have $t = 51,363$ s or 14.3 h. It is interesting to note the relative simplicity of solving Eq. (5.21) when the conversion, X, is given. If we have to find the conversion level of the substrate when time is given, it becomes more challenging because of the nonlinear nature of Eq. (5.21). It would be illustrative to use a software package to find the level of conversion reached after, say, 7 h of reaction.

5.5 Kinetics of Microbial Reactions

A general cellular reaction involves four classes of components: (1) *Substrate* – present outside the cell and which can be metabolised, (2) *Metabolic product* – produced by the cell which is excreted outside, (3) *Biomass constituents* and (4) *Intracellular metabolites*. It is difficult to distinguish clearly between *biomass constituent* and *intercellular metabolites*. The differentiation is normally based on the time scale of their turn-over in reactions. If the turnover is relatively rapid, it is treated as an intracellular metabolite; otherwise, it is simply included as a part of the biomass. Normally, an unstructured approach is taken to represent a cellular process, similar to the equation described in Chap. 1 (Sect. 1.3.2), where the biomass is approximated by a chemical formula determined by the proportion of various elements present in it. In general, if the substrates are termed S_i; metabolic products – P_i; and biomass constituents – X_i, the stoichiometric representation of the cellular reaction is:

$$\sum_{i=1}^{N} \alpha_i S_i + \sum_{i=1}^{M} \beta_i P_i + \sum_{i=1}^{Q} \gamma_i X_i = 0$$

where α_i, β_i and γ_i are the stoichiometric coefficients, which will be negative if the component is consumed, and positive if the component is formed.

Broadly speaking, there are two types of models used for describing the kinetics of microbial reactions: (1) *Structured growth models*: These models consider individual or groups of reactions occurring within cells (e.g. those involving DNA, RNA, Proteins etc in metabolic pathways), and (2) *Unstructured growth models*: In these models, a cell is considered to be a single unit interacting with the environment. Such models are simple to apply, and will be considered here.

One of the earliest unstructured growth model, known as the Malthus model, hypothesises that the rate of increase of dry cell weight per unit volume is proportional to the dry cell weight itself. In other words, if x is the dry cell weight per unit volume,

$$\frac{dx}{dt} = \mu x \tag{5.22}$$

where μ is known as the specific cell growth rate (s^{-1}). If μ is treated constant, this model predicts unlimited cell growth, which does not seem realistic! In 1942, Monod presented a model which linked the specific growth rate, μ, to the substrate concentration, S. Thus,

$$\mu = \left(\frac{\mu_{max} S}{K_s + S} \right) \tag{5.23}$$

where μ_{max} is the maximum specific cell growth rate. It is interesting to note the similarity between Eqs. (5.23) and (5.19) for enzyme kinetics, where k_M, the Michaelis constant, and K_s, the Monod constant, are analogous terms. Following the same logic as explained in the paragraph below Eq. (5.19), K_s can be shown to be the limiting nutrient concentration (S) resulting in half the maximum growth rate. Typical values of the μ_{max} range from 0.01 to around 3 h^{-1}, whereas K_s is typically 0.1 kg m^{-3}. Substituting the expression for μ from Eq. (5.23) into Eq. (5.22), the Monod model for cell growth may be written as:

$$\frac{dx}{dt} = \mu x = \left(\frac{\mu_{max} S}{K_s + S} \right) x \tag{5.24}$$

The limiting forms of Monod equation can now be developed. When $S \gg K_s$, then $(K_s + S) \approx S$, and the specific growth rate is independent of S. Equation (5.24) can be integrated with the initial condition, $x = x_0$ when $t = 0$ (i.e. the inoculum concentration) to yield:

$$\ln \left(\frac{x}{x_0} \right) = \mu_{max} t \quad \text{or} \quad x = x_0 \exp (\mu_{max} t) \tag{5.25}$$

This condition describes the exponential growth phase, which is observed in the initial stages of a fermentation reaction. When $S \ll K_s$, then $(K_s + S) \approx K_s$ and the specific growth rate is given by

$$\mu = \left(\frac{\mu_{max}}{K_s}\right) S \tag{5.26}$$

i.e. the specific growth rate is proportional to S. It is also interesting to note that a plot of μ versus S will be similar to Fig. 5.4.

Nutrient availability has a major influence on the specific growth rate. If a nutrient is available in concentrations that limit the growth of the cells, then that nutrient is termed a *growth limiting nutrient*. When the growth limiting nutrient is a major carbon and energy source, then it is referred to as a *growth limiting substrate*. Some fermentation media are designed such that, at the end of the fermentation, only the carbon source is in limiting supply. Others are designed such that nitrogen availability is growth limiting. Examples of such fermentations are the production of Polyhydroxybutyrate (PHB) and algal pigments. In some fermentations, the availability nutrients other than the carbon and energy source is intentionally controlled so as to slow or even prevent cell growth. This is done to ensure that the cells catalyse the conversion of a substrate to a product and not use the substrate for the synthesis of cellular building blocks. In large aerobic fermentation systems O_2 availability will inevitably be rate limiting for much of the fermentation, but this problem will be separately discussed in Sect. 5.6. Sometimes, high substrate concentrations can inhibit microbial growth by a variety of mechanisms. Monod equation, i.e. Eq. (5.23) can be modified to account for substrate inhibition by introducing an inhibition coefficient, K_I:

$$\mu = \mu_{max} \left(\frac{S}{K_S + S + \left(S^2/K_I\right)}\right) \tag{5.27}$$

The above discussion has focussed on the relationships between the cell growth rate (μ), biomass concentration (x) and substrate concentrations (S). The product formation rates, on the other hand, may be related to biomass and substrate concentration changes, but it is not necessary. The product formation rates can be *associated with cell growth*, e.g. in the case of ethanol production by glucose fermentation, *or not be* associated with cell growth, e.g. antibiotics, vitamins etc. which are secondary metabolites produced at the end of the exponential growth phase. When product formation is associated with cell growth, it is possible to define a yield coefficient linking the amount of product formed to the amount of substrate consumed and/or biomass generated. This enables the amount of product formed to be estimated, once biomass produced or substrate consumption has been evaluated on the basis of Monod equation. Suppose Y_{xS} is the yield coefficient of the biomass with respect to the substrate, i.e. the ratio of the amount of biomass formed to the amount of substrate consumed, the instantaneous relationship between substrate and cell masses can be written as:

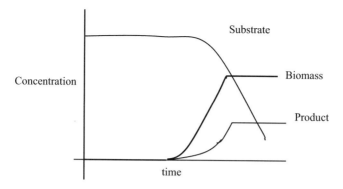

Fig. 5.5 Concentration profiles of substrate, biomass and product based on Monod model

$$\frac{dx}{dt} = -Y_{xS}\frac{dS}{dt} \tag{5.28}$$

The transient substrate concentration during fermentation can then be written as:

$$S = S_0 - \frac{1}{Y_{xS}}(x - x_0) \tag{5.29}$$

where S_0 and x_0 are the initial substrate and biomass concentrations. Now that we have a relationship between substrate and cell concentrations, the expression for S given by Eq. (5.29) can be substituted in the Monod model (Eq. 5.24) and the resulting differential equation can be solved to yield:

$$A \ln\left(\frac{x}{x_0}\right) - B \ln\left(\frac{Y_{xS}S_0 + x_0 - x}{Y_{xS}S_0}\right) = \mu_m t \tag{5.30}$$

where $A = \frac{K_S Y_{xS} + Y_{xS}S_0 + x_0}{Y_{xS}S_0 + x_0}$ and $B = \frac{K_S Y_{xS}}{Y_{xS}S_0 + x_0}$

In the exponential growth phase, x/x_0 increases rapidly, by S/S_0 hardly changes. Subsequently, S/S_0 drops sharply (Fig. 5.5).

5.6 Progress of Microbial Reactions in Batch and Continuous Processes

Very broadly speaking, microbial fermentations can be carried out either in a *batch* or a *continuous* manner. In batch processes, all the substrates, except oxygen (or air) are loaded into a reactor, after which the inoculum is added and the fermentation is allowed to progress. The biomass mass balance is given by Eq. (5.24), the substrate mass balance is given by Eq. (5.29). The theory and models developed in Sect. 5.5

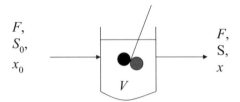

Fig. 5.6 Schematic representation of a continuous fermenter (volume V m3) with a feed entering at F m³s⁻¹ containing substrate at a concentration of S_0 kg m⁻³ and biomass concentration of x_0 kg m⁻³. S and x are the exit concentrations of concentrations of the substrate and biomass

are indeed valid for a batch process. In a continuous process, the substrate streams enter the reactor continuously and product streams are also withdrawn continuously at the same rate, so that the reaction rates in the reactor are maintained constant. The theory of continuous fermentations will now be developed.

5.6.1 Analysis of Continuous Stirred Fermenter or Chemostat

A steady state mass balance for the biomass can be written on the basis that the rate of biomass entering must be equal to the rate of biomass leaving plus rate of biomass formed in the reactor (Fig. 5.6). Thus,

$$Fx = Fx_0 + r_x V \qquad (5.31)$$

Where r_x is the rate of biomass formed per unit reactor volume and V is the reactor Volume.

Likewise, for the substrate, the mass balance equation can be written as:

$$FS_0 = FS + \frac{F}{Y_{xS}}[x - x_0] \qquad (5.32)$$

If we divide each term in Eq. (5.31) by F, we have:

$$x = x_0 + r_x \left(\frac{V}{F}\right)$$

The term V/F represents the mean residence time of the fluid in the reactor (τ). The reciprocal of τ is denoted by D and it is known as the *dilution rate* (s⁻¹). Thus Eq. (5.31) can be expressed in terms as D as follows:

$$D = D\frac{x_0}{x} + \frac{r_x}{x} = D\frac{x_0}{x} + \mu \tag{5.33}$$

It may be noted that the term μ appears in the above equation because $r_x = \mu x$ according to Monod kinetic model. Generally, there is no biomass entering the reactor continuously because cell growth is an autocatalytic process, and it is possible to sustain growth without a continuous supply of cells, which implies that $x_0 = 0$. Equation (5.33), therefore, further simplifies to:

$$D = \mu = \frac{\mu_{max}S}{K_S + S} \tag{5.34}$$

It follows that the outlet substrate concentration is therefore:

$$S = \frac{DK_S}{\mu_{max} - D} \tag{5.35}$$

And the outlet cell concentration is therefore given by:

$$x = Y_{xS}(S_0 - S) = Y_{xS}\left[S_0 - \frac{DK_S}{\mu_{max} - D}\right] \tag{5.36}$$

As $D \to 0$, (i.e. very high residence time), the substrate is completely consumed in the reactor and $S \to 0$, which in turn suggests that the outlet cell concentration, $x \to Y_{xS}S_0$. In other words, there is complete conversion of substrate into cells. There is also an upper limit to D, denoted by D^*, which occurs when x = 0 (i.e. when no cells leave the reactor or are present in it). Substituting $x = 0$ in Eq. (5.36), we have:

$$D^* = \frac{\mu_{max}S_0}{K_S + S_0} \tag{5.37}$$

For many systems, $S_0 \gg K_s$, so $D^* = \mu_{max}$. This is the *washout point*, where the rate of removal of cell exceeds the maximum rate of production in the fermenter. Two key features of a continuous fermenter are worth summarising: The *exit substrate concentration is independent of feed substrate* concentration and only depends on dilution and growth kinetic parameters. Secondly, the exit cell concentration depends on S_0.

The *productivity* of a continuous fermenter is defined as the cell production rate per unit volume. Thus, the productivity is given by:

$$P = r_x = Dx = DY_{xS}\left(S_0 - \frac{DK_S}{\mu_{max} - D}\right) \tag{5.38}$$

For maximum productivity, $(dP/dx) = 0$, which yields:

$$D_{opt} = \mu_{max}\left[1 - \sqrt{\frac{K_S}{S_0 + K_S}}\right]$$ (5.39)

Further, the optimum cell concentration is:

$$x_{opt} = Y_{xS}\left[S_0 + K_S - \sqrt{K_S(S_0 + K_S)}\right]$$ (5.40)

In practice, $S_0 \gg K_s$, $\therefore D_{opt} \approx \mu_{max}$; and $x_{opt} \approx Y_{xS} S_0$, and maximum productivity,

$$P = D_{opt}Y_{xS}S_0$$ (5.41)

Problem 5.4 **The growth of baker's yeast on glucose in a batch reactor is described by:**

$$C_6 H_{12} O_6 + 3O_2 + 0.48 NH_3 \rightarrow 2.88 CH_{1.67} N_{0.167} O_{0.5} + 4.32 H_2O + 3.12 CO_2$$

It is proposed to grow yeast in a fermenter, volume 1m³, with an initial glucose concentration of 100 kg m⁻³, and an inoculum concentration of 2 kg m⁻³. The yeast growth kinetics with glucose as limiting substrate is given by:

$$\mu = \frac{0.2\ S}{(0.1\ +\ S)} \qquad (h^{-1})$$

where μ is the specific growth rate (kg m⁻³ h⁻¹) and S is the substrate concentration. If the fermentation time is 14 h, would an assumption of exponential growth be reasonable on the basis of this information? Estimate the cell and residual substrate concentrations at the end of the batch process.

Solution The initial substrate concentration $S_0 = 100$ kg m⁻³, and the value of the Monod constant from the rate equation given is $K_s = 0.1$ kg m⁻³. Clearly, $S \gg K_s$ and therefore it is reasonable to assume that $\mu = 0.2$ kgm⁻³ h⁻¹. Therefore Eq. (5.25) is valid, and the biomass concentration after 14 h can be estimated from: $x = x_0 exp$ (μt), with $x_0 = 2$ kg m⁻³, to yield: $x = 32.9$ kg m⁻³. Therefore, the total biomass formed per cubic meter in 14 h = (32.9–2) = 30.9 kg m⁻³.

The residual substrate concentration can be estimated by knowing the yield coefficient of the biomass w.r.t. glucose. From the biochemical equation given above,

Y_{xs} = 2.88(molecular weight of the biomass)/(mol weight of glucose) = 2.88 × 23.95/180 = 0.383 kg biomass/kg glucose. Therefore the mass of glucose consumed is 30.9/0.383 = 80.64 kg m⁻³, which leaves a final substrate concentration of (100–80.64) = 19.36 kg m⁻³, which is still significantly greater than K_s. Thus, an assumption of exponential growth rate seems reasonable throughout the fermentation.

Problem 5.5 **A 60 m³ continuous stirred tank fermenter is to be operated with a sterile feed containing 12 kg m⁻³ of glucose for the synthesis of ethanol by**

Zymomonas mobilis. The concentration of glucose in the exit stream is to be reduced to 1.5 kg m^{-3}. The growth of the organism with glucose as the limiting nutrient can be described by the Monod equation with $\mu_{max} = 0.3$ h^{-1} and $K_s = 0.2$ kg m^{-3}. Studies on stoichiometry indicate that the growth in the above fermenter can be represented as follows:

  1 kg glucose → 0.06 kg cell + 0.46 kg ethanol + other products

  Estimate the dilution rate at which the fermenter should be operated, the concentration of the biomass and ethanol in the product stream, and the overall rate of production of ethanol.

Solution According to Eq. (5.34), $D = \mu = \frac{\mu_{max}S}{K_S+S}$ for a continuous fermentation system. In this problem, $\mu_{max} = 0.3$ h^{-1}, $K_s = 0.2$ kg m^{-3}, $S = 1.5$ kg m^{-3}. Substituting these values in Eq. (5.34), $D = 0.265$ h^{-1}. The concentration drop of substrate (glucose) in the reactor is: $(S_0 - S)$ kg m^{-3}. Given that the yield coefficient of biomass w.r.t glucose is 0.06, the concentration of biomass formed is: $0.06(S_0 - S)$ kg m^{-3} h^{-1} $= 0.06(12 - 1.5) = 0.63$ kg m^{-3}. The yield coefficient of ethanol w.r.t glucose is 0.46. Therefore the concentration of ethanol in the exit stream is: 0.46 $(12 - 1.5) = 4.83$ kg m^{-3}. The overall productivity of ethanol $=$ dilution $\times$ concentration of ethanol $= 1.28$ kg m^{-3} h^{-1}.

5.7 Multiphase Reactions: Combining Mass Transfer with Reaction Kinetics

In Problem 5.4, the growth kinetics of yeast has been modelled with glucose as the limiting substrate. In practice, the limiting substrate is more likely to be oxygen, because the oxygen has to transfer from air that is sparged into the liquid phase and then participate in the reaction in a dissolved state. Thus, there are two steps involved in such situations: (1) Mass transfer from the gas to the liquid state, followed by (2) reaction of dissolved oxygen with other substrates. Both these steps occur in series, and the overall reaction rate will be determined by the slower of the two steps – which, in practice, is the mass transfer step. In Chap. 4, Sect. 4.7, the rate of oxygen transfer from air into a liquid medium is given by Eq. (4.35): $r = k_L a$ $(C^* - C_0)$ where r is the rate of oxygen transfer per unit volume of the reaction mixture (kg m^{-3} s^{-1}), C^* is the saturation solubility of oxygen in the liquid medium and C_0 is the concentration of dissolved oxygen. The maximum rate at which oxygen can be transferred is $k_L a \, C^*$. The rate at which oxygen is consumed in the reaction can be obtained by using the Monod model for biomass growth (i.e. $v = \mu x$) and introducing a yield coefficient for biomass w.r.t oxygen, Y_{XO}. The maximum rate at which oxygen can be consumed is therefore $\mu_{max}x/Y_{XO}$. The ratio of the maximum rate of substrate consumption to the maximum rate of substrate transfer is known as the Damkohler number:

$$Da = \frac{\mu_{max}x}{Y_{XO}k_La C^*} \tag{5.42}$$

A relatively low value of Da << 1 implies that the reaction rate is slow; so reaction rate, not mass transfer, controls the rate of the overall process. A high value of Da >> 1 implies that the reaction rate is quite fast and mass transfer rate controls the overall process. A mass balance for oxygen can be written as follows:

$$\frac{dC_0}{dt} = k_La(C^* - C_0) - \frac{\mu x}{Y_{XO}} \tag{5.43}$$

Fermentation reactions are generally carried out by maintaining a constant level of dissolved oxygen. In other words, $dC_0/dt = 0$ in Eq. (5.43), which leads to the equating of the rates of oxygen transfer and consumption, i.e.

$$k_La(C^* - C_0) = \frac{\mu x}{Y_{XO}} \tag{5.44}$$

It is also necessary to note that aerobic cells also need a minimum concentration of dissolved oxygen below which the oxygen concentration will limit the rate of cell respiration. This concentration is known as the *critical oxygen concentration*. It is customary to operate a reactor with the dissolved oxygen level greater than this value. The concentration of dissolved oxygen in a fermentation medium can be obtained either from Eqs. (5.43) or (5.44) depending on whether the process is under unsteady state conditions or a quasi-steady state.

Problem 5.6 A bioreactor has an oxygen mass transfer coefficient capability of 400 h^{-1}. What is the maximum concentration of E. coli that can be grown aerobically in this reactor. Respiration rate of *E. coli* is 0.35 g O2 (g Cell)$^{-1}$ h^{-1}. Critical oxygen concentration is 0.2 mg/L. Assume oxygen saturation with air to be 6.7 mg/L.

Solution The rate of oxygen transfer limits the growth of *E.Coli* in the reactor. Therefore, the maximum concentration of *E coli* can be grown under the conditions of maximum rate of oxygen transfer, which in turn will occur when the driving force $(C^* - C_0)$ is maximum. Clearly C^* is determined by the thermodynamic solubility of oxygen (6.7 mg L^{-1}) and cannot be changed. The maximum driving force will occur when the level of dissolved oxygen is maintained at its lowest value possible, i.e. at the critical concentration (0.2 mg L^{-1}). The maximum oxygen transfer rate $= k_La(C^* - C_0) = 400(6.7-0.2)$ mg L^{-1} h^{-1}. Since the respiration rate is 0.35 g O$_2$ (g Cell)$^{-1}$ h^{-1} or 350 mg O$_2$ (g Cell)$^{-1}$ h^{-1}, the biomass concentration is: $\frac{400 \times 6.5}{350} = 7.43$ mg L^{-1}.

Problem 5.7 Yeast is grown aerobically in batch culture on a glucose feed with concentration 60 kg m^{-3}; the growth yield coefficient is 0.5 kg kg^{-1}. The maximum value of the volumetric oxygen transfer coefficient k_La is 0.125 s^{-1} and it is intended that the dissolved oxygen concentration should not fall below

0.005 kg m^{-3}. It may be assumed that growth kinetics are always oxygen controlled, with

$$\mu = \frac{0.25 C_o}{0.001 + C_o} \qquad \left(h^{-1} \right)$$

Would the oxygen transfer in the system be adequate (a) when $x = 5$ kg m^{-3} and (b) towards the end of the fermentation? Estimate the dissolved oxygen concentration and specific growth rate under condition (b). Note that the solubility of oxygen in the medium $C^* = 0.01$ kg m^{-3} and the yield coefficient of biomass w.r.t. oxygen is: $Y_{02} = 5$ kg kg^{-1}.

Case (a): the biomass concentration x $= 5$ kg m^{-3}. If the dissolved oxygen concentration is maintained at its lowest threshold, $C_0 = 0.005$ kg m^{-3}. The specific biomass growth rate

$$\mu = \frac{0.25 \times 0.005}{0.001 + 0.005} = 0.21 \ h^{-1} \quad \text{and the rate of growth of bio-}$$

mass $= \mu x = 1.05$ kg m^{-3} h^{-1}. The rate at which oxygen must be consumed to support this biomass growth rate $= \mu x / Y_{02} = 1.05/5 = 0.21$ kg m^{-3} h$^{-1} = 5.83 \times 10^{-5}$ kg m^{-3} s^{-1}. The oxygen transfer rate $= k_L a (C^* - C_0) = 0.125(0.01-0.005) = 6.25 \times 10^{-4}$ kg m^{-3} s^{-1}. Clearly the rate of oxygen transfer is an order of magnitude greater than the rate of oxygen consumption, and it is more than adequate.

Case (b): Towards the end of fermentation, it would be reasonable to assume that all glucose has converted to biomass. Since the starting concentration of glucose is 60 kg m^{-3}, and the yield coefficient w.r.t. glucose is 0.5 kg kg^{-1}, the biomass concentration will be $x = 30$ kg m^{-3}. The biomass growth rate will be $\mu x =$

$\dfrac{0.25 C_o}{0.001 + C_o} \times 30$ and the oxygen consumption rate will be $\dfrac{0.25 C_o}{0.001 + C_o} \times \frac{30}{5}$ kg

m^{-3} h$^{-1} = \dfrac{0.25 C_o}{0.001 + C_o} = \frac{4.17 \times 10^{-4} C_o}{0.001 + C_o}$ kg m^{-3} s^{-1}. The rate of oxygen transfer is:

$k_L a (0.01 - C_0)$ kg m^{-3} s^{-1}. Assuming that the rate of oxygen transfer is equal to the rate of oxygen consumption, we have:

$k_L a (0.01 - C_0) = \frac{4.17 \times 10^{-4} C_o}{0.001 + C_o}$, which is a quadratic equation in C_o and can be

solved to yield: $C_0 = 0.007$ kg m^{-3}. The specific growth rate $\mu =$

$\dfrac{0.25 \times 0.007}{0.001 + 0.007} = 0.22 \ h^{-1}$.

A second example involving mass transfer with biochemical reactions arises when the enzyme is immobilised, i.e. fixed to a support. When enzymes are used in the homogeneous liquid state, they often get washed out with the product and their isolation for reuse can be difficult and quite expensive. One way of avoiding this situation is to fix the enzyme on a solid support. Although this approach can avoid

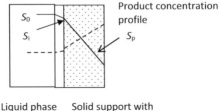

Liquid phase   Solid support with
                 enzyme immobilised

Fig. 5.7 Immobilised enzyme system showing substrate concentration profile where S_0 is the bulk liquid concentration, S_i is the substrate concentration at the support-liquid interface, and S_p is the substrate concentration inside the support, which also follows a gradient. There are two mass transfer resistances: liquid side mass transfer resistance and intra-particle mass transfer resistance. The product is formed in the support and has to diffuse out, and its concentration profile is shown by the dash line

enzyme loses and result in better enzyme utilisation, the substrate has to diffuse from the bulk solution into the enzyme support where the reaction can occur. The product must then diffuse back into the bulk solution. The kinetics of such reactions can be described by the Michaelis Menten equation, but the substrate concentration to be used in the model will be the value prevailing in the vicinity of the enzyme support; not the bulk concentration. Further, since mass transfer principles determine the rate at which the substrate can diffuse from the bulk solution into the support structure, there must exist a concentration driving force which causes the substrate concentration in the vicinity of the enzyme to be less than the concentration in the bulk. This can potentially slow down reaction rates significantly.

In practice, one can envisage at least two mass transfers and one reaction step constituting such reactions, as shown in Fig. 5.7

A relatively straightforward situation to model is one where the enzyme is evenly distributed on a nonporous support material, all enzyme molecules are equally active, and substrate diffuses through a thin external liquid film to reach the reaction sites. We can consider all rates to be based per unit area of the solid liquid interface. Similar to Eq. (5.42), we can define a Damkohler number as the ratio of the maximum reaction rate to the maximum mass transfer rate:

$$\mathrm{Da} = \frac{v'_{max}}{k_L S_0} \qquad (5.45)$$

Note that v'_{max} denotes the maximum rate of enzyme catalysed conversion per unit area. Again, if $\mathrm{Da} \gg 1$, the reaction rate is high, so mass transfer controls the overall rate, and *vice versa*.

In general, if S_p is the substrate concentration on the particle and S_0 is the bulk liquid concentration, we can assume a quasi-steady state and equate the flux of substrate diffusion to the amount of reaction taking place per unit area, just as Eq. (5.44):

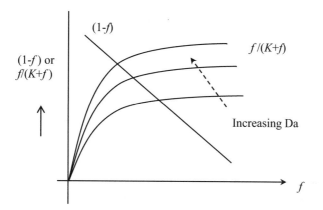

Fig. 5.8 Graphical solution of Eq. (5.36) which enable the estimation of f for different Damkohler numbers and Michaelis Menten constants. As Da increases, the f value at the intersection point decreases and $S_p \to 0$ indicating that the external mass transfer resistance increases in relation to the reaction kinetics. When the external mass transfer completely dominates, $S_p = 0$ and the rate of the reaction is simply $v = k_L\, S_0$. In other words, the reaction rate is independent of the intrinsic Michaelis Menten parameters

$$k_L\left(S_0 - S_p\right) = \frac{v'_{\max} S_p}{k_M + S_p} \tag{5.46}$$

Even though Eq. (5.46) is a quadratic equation in S_p and can be analytically solved, it would be best to normalise the variables using: $(S_p/S_0) = f$, $(k_M/S_0) = K$ and Eq. (5.34) for Da to yield:

$$1 - f = \mathrm{Da}\left(\frac{f}{K + f}\right) \tag{5.47}$$

The graphical solution of Eq. (5.47) is essentially the point of intersection of the plot of $(1 - f)$ against f and $f/(K + f)$ against f for given values of Da; a trend plot is shown in Fig. 5.8

The *effectiveness factor*, η, is another term which is extensively used in the discussion on immobilised enzyme systems and it represents the ratio of the actual reaction rate under immobilised conditions to that had there been no mass transfer limitations:

$$\eta = \left(\frac{S}{k_M + S_p}\right) \Big/ \left(\frac{S_0}{k_M + S_0}\right) \tag{5.48}$$

Fig. 5.9 Reaction rate data
for two enzyme catalysed
reactions with the enzymes
immobilised on a
non-porous support

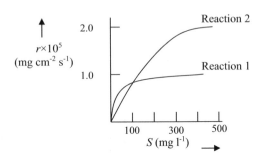

Problem 5.8 **The reaction rates for two different enzyme catalyzed reactions, both involving enzyme immobilization onto a non-porous support material, is shown in** Fig. 5.9 **below. Estimate the approximate values of the mass transfer coefficient k_L for the two reactions.**

Solution It is reasonable to assume that the reaction will be mass transfer limited when the bulk substrate concentration is low; this is because the Damkohler number (Eq. 5.45) will be high. The concentration on the immobilized surface will be virtually zero. Therefore, the rate of the reaction is $k_L(S_0 - S_p) \approx k_L S_0$. In other words, k_L can be approximated to the initial slope of the graph shown in Fig. 5.9. For reaction 1: $k_L = 1.0 \times 10^{-5}/(50 \times 10^{-3}) = 2.0 \times 10^{-4}$ cm/s. Similarly, for reaction 2: $k_L = 9.5 \times 10^{-5}$ cm/s.

Note that these rates become reaction rate limited at high substrate concentrations and r becomes equal to r_{max}. But it is important to note that the r_{max} values evident in the figure *do not* represent the intrinsic Michaelis Menton parameter, because of mass transfer resistance.

Further Reading

Doran PM (2013) Bioprocess engineering principles, 2nd edn. Academic Press. ISBN-13: 978-0-12-220851-5.
Ravi R, Vinu R, Gummadi SN (2017) Coulson and Richardson's chemical engineering. Volume 3A: chemical & biochemical reactors and reaction engineering, 4th edn. Elsevier. (Chapters 1 and 6)

Chapter 6
Phase and Reaction Equilibrium, and Phase Transitions

Aim The aim of this chapter is to illustrate how phase and reaction equilibrium concepts are applied in food processes and products to understand water activity, product stability, phase transitions and separation processes such as adsorption. Background knowledge of basic thermodynamics such as first and second laws, enthalpy and entropy, and free energy functions is a pre-requisite to follow the contents of this chapter.

6.1 Introduction

Thermodynamic functions and the relationships between these functions provide considerable insights into: (1) the effects of thermal and other processing methods on foods as well as (2) the physical and biochemical transitions occurring in foods during longer term storage under various conditions. The key thermodynamic functions relevant to foods will be first considered.

6.1.1 Enthalpy (ΔH)

Enthalpy (ΔH) is essentially represents the heat content of a system with respect to a defined baseline (as explained in Chap. 1, Sect. 1.4). *Enthalpy of formation* (ΔH_f) of a chemical or biochemical is the heat involved in forming a mole of the chemical under standard state conditions (1 atmosphere and 298 K) starting from the elements constituting it under the same conditions. Enthalpy of a reaction (ΔH_R) is essentially the difference between the enthalpies of formation of the products and those of the reactants, taken under the reaction conditions, normally expressed per mole of a product or a reactant. A spontaneous process moves in the direction of lowering enthalpy, i.e. $\Delta H < 0$.

© Springer Nature Switzerland AG 2022
K. Niranjan, *Engineering Principles for Food Process and Product Realization*,
Food Engineering Series, https://doi.org/10.1007/978-3-031-07570-4_6

6.1.2 Entropy (S)

Entropy (S), based on classical thermodynamics, is a property of the state of a material like pressure, temperature and internal energy. The second law of thermodynamics also states that all the internal energy associated with any material cannot be converted to useful work, which implies that a part of this energy is "unavailable". Entropy change is defined as this "unavailable" energy divided by the absolute temperature; i.e. $\Delta S = Q/T$. The reason why this energy is unavailable to do useful work is because it is associated with structural conformation, and represents the "disorder" within the system. The unit of ΔS is J K^{-1}. The entropy change in a spontaneous process is positive. In other words, spontaneity changes a system from a more to less orderly state (e.g. ice is very well structured, but melting it results in greater entropy and disorder). This ΔS is positive or >0 for a spontaneous process.

6.1.3 Gibbs Free Energy (G)

Gibbs free energy (G) is defined as $G = H - TS$. If H represents the system enthalpy and TS represents the energy associated with its structural conformation, $(H - TS)$ represents the energy that is *free* or *available* for it to undergo changes. For spontaneous process, $\Delta H < 0$ and $\Delta S > 0$. Hence $\Delta G < 0$, which a good test to diagnose the spontaneity of a process. If a process is endothermic, i.e. $\Delta H > 0$ and $\Delta S > 0$, it can be spontaneous if $(\Delta H - \Delta TS)$ is negative, i.e. if $T > (\Delta H/\Delta S)$. Likewise. if $\Delta H > 0$ and $\Delta S < 0$, the process can only be spontaneous if $T < (\Delta H/\Delta S)$. Further, at equilibrium, $\Delta G = 0$. Based on the standard conditions defined in Sect. 6.1.1, we can also define the standard Gibbs free energy change ΔG^0 for a chemical reaction, which can be shown to be given by:

$$\Delta G^0 = -RT \ln K \tag{6.1}$$

where K is the equilibrium constant.

6.2 Water Activity of Foods (a_w)

Water is the main, and arguably, the most important constituent of foods. The water content of food influences the stability, quality and safety of foods, and can effect and affect phase transitions, microbial growth, enzymatic and non-enzymatic reactions, lipid oxidation and texture. However, the properties of food, especially its transient properties, are not directly related to the total water or moisture content. This is because, water exists in different states within a given food, and it also associates in different ways with the other food components, both of which can be critical in determining the properties and the fate of the food. Therefore, the activity of water can be quite different to the moisture content per se, and the relationship between the two can be quite complex. For all practical purposes, the water activity is defined as the ratio of the partial pressure of water around a food when in a completely undisturbed balance with the surrounding air (p_w) and the vapor pressure of distilled water at the same temperature (p_{w*}). Thus,

$$a_w = \frac{p_w}{p_w^*} \tag{6.2}$$

Based on the definition given in Eq. (6.2), a water activity of 0.85 means the partial pressure due to the food is 85% of that of pure water. Most foods have a water activity above 0.95 which will provide sufficient moisture to support the growth of bacteria, yeasts, and mould. Products that lie in the water activity range of 0.60–0.90 are considered to be intermediate moisture products, and those having water activities greater than 0.90 are high water activity products. Products with $a_w < 0.6$ are generally considered to be dry.

The water activity isotherm for a typical low moisture food is shown in Fig. 6.1. The figure also shows the existence of hysteresis which is observed when experiments are done by progressively wetting the product (sorption) or progressively drying a sample from an initially wet state (desorption). The isotherm is generally higher during desorption than during sorption. The reasons for the observed hysteresis are not entirely conclusive and a critical account of various reasons is given by Caurie (2007).

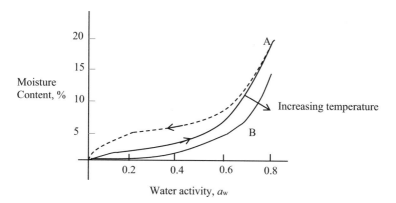

Fig. 6.1 Typical plot of moisture content against water activity for low moisture foods at a given temperature. In respect of curves A, the solid line represents moisture sorption while the dashed line represents desorption curve indicating hysteresis. Curve B represents the isotherm at a higher temperature

6.2.1 *Measurement of Sorption Isotherms*

There are three methods commonly used to construct sorption isotherms.

1. *Traditional desiccator method*: In this method, each point on the isotherm is determined by equilibrating a sample to a known water activity and then determining its equilibrium moisture content by determining the weight. Typically, the sample is placed in a sealed laboratory desiccator over a saturated salt slurry. Different water activity levels can be achieved by using different salt solutions; see Table 6.1. The equilibration process can take weeks. It is therefore necessary to control the temperature tightly, especially to prevent microbial contamination which can occur when $a_w > 0.60$

2. *Dynamic vapour sorption method*

   Automated instruments, often referred to as controlled atmosphere balances, hold the sample at one water activity level until the sample weight stops changing and remains constant., The instrument then measures the water content by weight, and then dynamically moves to the next water activity which is achieved by controlled by mixing dry and wet air. Such automatic isotherm generators are much faster and less labour intensive, and these instruments can also be used to determine sorption kinetics.

3. *Dynamic dew point isotherm method*

   In this method, adsorption occurs as saturated wet air is passed over a sample. The airflow is stopped after which the water activity and the weight are directly measured. The sample does not have to wait for equilibration to a known water activity, so this method is faster and is able to result in considerable data in a relatively short space of time.

Table 6.1 Saturated solution of salts and their water activity at 25 °C (Pollio et al. 1996)

Saturated salt solution	a_w
LiCl	0.1105
$MgCl_2$	0.330
NaBr	0.577
$CoCl_2$	0.633
$SrCl_2$	0.712
NaCl	0.751
NH_4Cl	0.772
$(NH4)_2SO_4$	0.803
KBr	0.809
KCl	0.842
$BaCl_2$	0.903
KNO_3	0.927
$CuSO_4$	0.969
K_2SO_4	0.975

6.2.2   Moisture Sorption Isotherm (Models)

It is estimated that there are around 270 models describing the relationship between moisture content and water activity. Only four of the models will be considered here just to illustrate the principles.

The simplest way to model the relationship between the moisture content M_w (kg water/kg solid) and water activity is to assume a linear relationship between the two:

$$M_w = B_{linear}a_w + C_{linear} \qquad (6.3)$$

where B_{linear} and C_{linear} are model constants (kg water (kg solids)$^{-1}$). Although the model is simple, it is only applicable over very narrow water activity ranges where a linear relationship can be assumed to prevail on the isotherm.

The Brunauer-Emmett-Teller (BET) model is a more realistic non-linear model, which has two model parameters: moisture content of the monolayer adsorbed ($M_{1, BET}$), and the Guggenheim constant for the BET model (C_{BET}), and can be written as:

$$\frac{M_w}{M_{1,BET}} = \frac{C_{BET}a_w}{(1 - a_w)(1 - a_w + C_{BET}a_w)} \qquad (6.4)$$

It may be noted that C_{BET} can be correlated with temperature as follows:

$$C_{BET} = e^{Q_S/RT} \qquad (6.5)$$

where Q_S is known as the surface interaction energy (J mol^{-1}), R is the ideal gas constant and T is the temperature in K. The BET equation is generally applicable to low moisture content foods and in the activity range $0.1 < a_w < 0.35$

The Guggenheim-Anderson-de Boer (GAB) model is an extension of BET model and fits sorption data up to $a_w = 0.9$. It is similar to BET model except that it has one more, i.e. three model parameters: moisture content of the monolayer adsorbed ($M_{0, GAB}$), the Guggenheim constant for the GAB model (C_{GAB}), and another constant K_{GAB}.

$$\frac{M_w}{M_{0,GAB}} = \frac{C_{GAB}K_{GAB}a_w}{(1 - K_{GAB}a_w)(1 - K_{GAB}a_w + C_{GAB}K_{GAB}a_w)} \qquad (6.6)$$

A relatively recent empirical model, called the Double Log Polynomial (DLP) model, has proven to be even better than the GAB at characterizing complex isotherms (Condon 2006). According to this model:

$$M_w = b_3\chi^3 + b_2\chi^2 + b_1\chi + b_0 \quad \text{where} \quad \chi = \ln\left[-\ln\left(a_w\right)\right] \qquad (6.7)$$

where b_0, b_1, b_2 and b_3 are the model constants.

The water activity *increases with temperature* at a given moisture content. Likewise, the amount of moisture that can be held at a given water activity goes down with increasing temperature (Fig. 6.1). The effect of temperature on water activity follows the well-known Clausius-Clapeyron equation which, in its integrated form, can be written as:

$$\ln\left(\frac{a_{w2}}{a_{w1}}\right) = \frac{\Delta H}{R}\left(\frac{1}{T_1} - \frac{1}{T_2}\right) \tag{6.8}$$

In Eq. (6.8) R (J mol^{-1} K^{-1}) is the gas constant and ΔH is the heat of sorption (J mol^{-1}) which is experimentally determined as the slope of the plot of ln a_w against $1/T$ at a given moisture content (see Arrhenius plot, Sect. 5.2.1).

Problem 6.1 A cake formulation requires the following: 60% cake (moisture content 15% wwb), 20% crème filling (moisture content 12% wwb) and 20% chocolate icing (moisture content 7% wwb). The moisture sorption isotherms are:

$$m_w = -9.8\ln\left[-\ln\left(a_w\right)\right] + 6.4 \text{ for the cake}$$
$$m_w = -0.5\ln\left[-\ln\left(a_w\right)\right] + 11.7 \text{ for the creme, and}$$
$$m_w = -2.9\ln\left[-\ln\left(a_w\right)\right] + 2.7 \text{ for the icing}$$

where the moisture content m_w is expressed as % on a wet weight basis. Estimate:

1. **The average moisture content of the formulation,**
2. **The water activity of each component before mixing, and**
3. **The equilibrated water activity after mixing**
4. **The final equilibrated moisture contents of each ingredient.**

Solution
1. The average moisture content is $= (0.6 \times 0.15 + 0.2 \times 0.12 + 0.2 \times 0.07) = 0.128$ or 12.8% wet weight basis
2. The initial water activity can be determined by plugging the initial moisture content in the respective isotherms. Thus for the cake $a_w = 0.66$; for the crème, $a_w = 0.58$; and for the icing, $a_w = 0.80$
3. The equilibrated water activity is essentially the value of a_w obtained, by taking the average of the right hand side of each individual isotherm; i.e. by solving the Equation: $0.6[-9.8\ln\left[-\ln\left(a_w\right)\right] + 6.4] + 0.2[-0.5\ln\left[-\ln\left(a_w\right)\right] + 11.7]$ $+0.2[-2.9\ln\left[-\ln\left(a_w\right)\right] + 2.7] = 12.8$ The value of $a_w = 0.67$.
4. The equilibrated moisture content of each component can be obtained, by putting $a_w = 0.67$ in each of the isotherms, to yield: Final moisture content of the cake: 15.5%; the final moisture content of crème $= 12.2\%$; and the final moisture content of the icing $= 5.4\%$. It is interesting to note that the cake and crème have marginally gain moisture after equilibrating, while the icing has lost moisture.

Thus, the moisture content of different components can be different at equilibrium, but the water activity is the same. In other words, moisture migrates to result in the same uniform water activity.

Problem 6.2 **A breakfast cereal bar with a moisture content of 25% is stored at 15 °C, a temperature at which it has a water activity of 0.6. The cereal bar suffers storage abuse and the temperatures increases to 40 °C. If the heat of sorption of the cereal bar is 3850 J (mol)$^{-1}$, estimate its water activity at this temperature and comment on its microbiological stability.**

Solution The water activity at the higher temperature can be directly determined from Eq. (6.8), where $a_{w1} = 0.6$, $T_1 = 288$ K, $T_2 = 313$ K, $\Delta H = 3850$ J (mol)$^{-1}$, and R = 8.34 J (mol)$^{-1}$ K^{-1}. Thus, $a_{w2} = 0.68$. At 15 °C, the water activity is 0.6 which is sufficiently low to prevent mould growth. However, following temperature abuse, its water activity approaches 0.7, which is quite capable of supporting mould growth. This implies that the rise in temperature can render the product microbiologically unstable and this can have food safety implications. Note that the moisture content of the product has not changed and remains the same at 25%

6.2.3 Kinetics and Equilibrium of Water Activity

Although water activity (a_w) is a property of a moist material which can be measured at any point in time, the material is invariably in contact with ambient air either directly or through a packaging barrier. The relative humidity of air (h_w), defined here as the ratio of the prevailing partial pressure of water in air to the vapour pressure at the same temperature, is different in value from the water activity. A driving force is therefore set up for the two water activities to equilibrate. If the air is very humid, moisture will be transferred to the food and its water activity can increase. Consider a low moisture food wrapped in a thin packaging film and placed in air with relative humidity h_w. Such a situation was considered in Chap. 4 and based on Eq. (4.9), the mass flux for water loss, can be written as:

$$\frac{dm_{H_2O}}{dt} = \frac{P}{L}A(p_1 - p_2) \tag{6.9}$$

In the above equation, the left hand side is the rate of water gain by the food (kg s^{-1}), P is the packaging permeability (Pa^{-1} m^{-1} s^{-1}kg), L is the packaging film thickness (m); A is the surface area for moisture transfer (m^2); p_1 is the partial pressure of water in air, and p_2 is the partial pressure of water in the air surrounding the food (both pressures in Pa).

If M (kg) is the mass of the food, we can assume that it does not change due to mass transfer and divide both sides of Eq. (6.13) by M. Note that (m_{H2O}/M) is its moisture content and can be replaced by its isotherm given by Eq. (6.3). Further,

$p_1 = h_w p^*$ and $p_2 = h_w p^*$ where p^* is the vapour pressure of water at the given temperature (Pa). Substituting these into Eq. (6.9), we get:

$$\frac{da_w}{dt} = \frac{PAp^*}{LMB_{lin}}(h_w - a_w) \tag{6.10}$$

The above equation indicates how the water activity can change with time. The reciprocal of the rate constant of Eq. (6.10) has time units and is known as the time constant of the process given by:

$$\tau = \frac{LMB_{lin}}{PAp^*} \tag{6.11}$$

The differential Eq. (6.10) can also be solved with the initial condition: $a_w = a_{wi}$ at $t = 0$, to give the water activity at any time:

$$a_w = h_w - (h_w - a_{wi})e^{-t/\tau} \tag{6.12}$$

In practice, the critical or threshold water activity of the product, beyond which the product will not be acceptable, is often known. For example, the product may not be microbiologically stable above $a_w = 0.6$ (as in Problem 6.2). Thus, Eq. (6.12) with Eq. (6.11) can be used to estimate the shelf life of products or deduce packaging film properties to store the product with integrity over a desired shelf-life period. It is also worth noting that a linear isotherm was assumed in solving the above problem, so that an analytical solution was possible. With software packages available, it is possible to do a similar exercise with non-linear and more complicated isotherms.

Problem 6.3 10 g of spray dried milk powder, $a_w = 0.1$, is packaged in a LDPE film of thickness 0.015 mm ad exposed to air having relative humidity of 0.6. If the area of the film exposed to mass transfer is 10 cm^2, the vapour pressure of water at 25 °C is 3000 Pa and the permeability of LDPE to water vapour is 5.1×10^{-16} kgm^{-1}s^{-1}Pa^{-1}, how long will it take for the water activity to increase to a critical value of 0.4? The linear isotherm slope for spray dried milk is 0.026 kg kg^{-1}.

The value of τ will first be calculated from Eq. (6.16), with $L = 1.5 \times 10^{-5}$ m, $M = 10^{-2}$ kg, $B_{lin} = 0.026$ kg kg^{-1}, $P = 5.1 \times 10^{-16}$ kgm^{-1}s^{-1}Pa^{-1}, $A = 10^{-3}$ m^2, $p^* = 3000$ Pa. The value of $\tau = 2.55 \times 10^6$ s (29.5 days). The time taken for $a_w = 0.4$ can be obtained from Eq. (6.17) with $a_{wi} - 0.1$, $h_w = 0.6$ and $\tau = 2.55 \times 10^6$ s, to give $t = 2.33 \times 10^6$ s (c.a. 27 days).

6.2.4 Water Activity and Phase Transitions

Moisture sorption isotherms, such as the one shown in Fig. 6.1 except with a much higher resolution, can be used to identify phase transition. The critical water activity at which the phase transition occurs can be identified as the point where the isotherm undergoes a sharp inflection; see Fig. 6.2. Such phase transitions can influence shelf life and product stability, and can also help identify the right processing regime. If the resolution of the isotherm is not high enough, it is quite possible that the critical water activities, and indeed the phase transition points are missed! The moisture sorption isotherm can also help identify whether a material is in the amorphous or crystalline material state, and provide information about the level of each in a product. This is illustrated in Fig. 6.2 by noting the isotherms for crystalline and amorphous sucrose. At higher temperatures, the critical water activities are generally lower, and a product can potentially lose stability without any change in water activity.

It may be noted that phase transition points can also be identified by determining what is known as *the glass transition temperature*, T_g, which represents the temperature at which an amorphous biopolymer, such as starch, changes from a hard, rigid or "glassy" state to a more pliable, or rubbery state. It may be noted that glass transition is typical of amorphous materials, whereas crystalline solids transition from solid to liquid state at the melting point. The glass transition temperature is practically determined using a *differential scanning calorimeter*, where a material is heated, i.e. temperature is scanned while holding its water activity constant. This is in contrast to sorption isotherms where the temperature is held constant as the water activity is scanned. Both, sorption isotherm as well as glass transition temperatures can identify key phase transitions in food systems. It is however arguable whether the water activity (a_w) or the state of the system as defined by the glass transition temperature (T_g) influences the rates of chemical and microbial reactions in low moisture content food systems. There are a number of research publications which discuss this aspect (e.g. Bell and Hageman 1994).

Fig. 6.2 Typical isotherms showing water activity at which phase transition is observed: deliquescence in crystalline sucrose at $a_w \cong 0.8$, phase transition in amorphous sucrose at $a_w \cong 0.3$, and phase transition in spray dried milk at $a_w \cong 0.45$

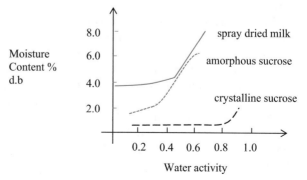

6.3   Equilibrium and Thermodynamics of Adsorptive Separation Processes

In Sect. 6.2.1, the term *sorption* was used to describe the transfer of water from humid air into foods. This term includes *adsorption* of water on the surface of the food, and subsequent transfer into the bulk, the cumulative effect being known as *absorption*. Thus sorption includes *adsorption* – which is a superficial process by which one material gets preferentially fixed to the surface of another, and *absorption* – which essentially occurs throughout the bulk of a material. In this section, we will exclusively consider the superficial process of *adsorption*, which has very wide applications in food processing, especially for separating components such as antioxidants and colour from foods and waste products. The component to be separated by adsorption is known as *adsorbate* and the material whose surface is used to attach the adsorbate is known as *adsorbent*. There are several adsorbents being used in practice for various applications which include natural materials as well as those which have been synthesised for specific applications. Increasingly, natural materials are being preferred for separating food components and adsorbents such as activated carbon, chitosan, and even dry fruit shells are being increasingly used.

There are many mechanisms by which an adsorbate attaches itself to an adsorbent. The most common force binding the two is of the van der Waal type. However, stronger forces – some as strong as chemical bonds like ionic or covalent bonds – are also possible, and processes such as ion exchange and chromatography involve such chemical interactions. When the binding between the two is very strong, the interaction between a combination of adsorbent and adsorbate can be highly specific. This specificity is exploited to selectively separate components such as proteins and enzymes in the biotechnology industry, where it is common place to exploit the *differences* between hydrophobicity, charges, molecular sizes of mixture components to separate and purify a targeted component. When such property differences become very subtle and selectivity diminishes, it is possible to find a ligand which will selectively bind to a targeted protein. In practice, this ligand is generally immobilised on a solid support and allowed to bind with the targeted protein for subsequent separation. This is the basis of *affinity based processes* such as affinity chromatography (Reichelt 2015).

The analysis of the kinetics and equilibrium aspects of chemical and affinity based interactions is highly complex and beyond the scope of this book. Here, we will consider the physical interactions between an adsorbent and an adsorbate. In its simplest state, the interaction is monomolecular, i.e. the adsorbate forms a single molecular layer on the adsorbent. Clearly, the maximum amount adsorbed at any given temperature and pH In the case of liquids (and pressure, if the adsorbate is a gas) depends on the area of the adsorbent and the molecular size of the adsorbate. Further, it is important to note that the *rate of adsorption* is a kinetic factor and *principles of mass transfer*, such as the models discussed in Chap. 5 can be applied, whereas the *maximum amount adsorbed* occurs at *equilibrium* and *principles of thermodynamics* can be applied. Just as we have sorption isotherms, we have adsorption isotherm which link the concentration of the adsorbate in a liquid with that on an adsorbent at a given temperature and pH. The simplest and most widely used adsorption isotherms are the *Freundlich* and *Langmuir isotherms*.

These models can be explained by considering the following situation. A volume V (m^3) of a solution (say, clear green tea extract containing polyphenols) with solute concentration C_0, kg m^{-3}) is taken in a beaker to which W kg of an adsorbent (say microporous starch granules) are added. The mixture is stirred and the solute (i.e. polyphenols) gets adsorbed on the starch granules. After a while, an equilibrium is reached when the concentration of the solute in the solution is C_e (kg m^{-3}) and the concentration of adsorbed polyphenols on the starch granules is Q_e (kg (kg adsorbent)$^{-1}$) – which is also known as the *adsorption capacity* of the adsorbent. Based on a mass balance,

$$Q_e = \frac{(C_0 - C_e)V}{W} \qquad (6.13)$$

The mass of the adsorbent per unit volume of the solution (i.e. W/V) is also known as the *adsorbent loading*, Γ. The product $(Q_e \cdot \Gamma) = q_e$ represents the concentration of the adsorbate on the adsorbent per unit volume of the solution, which is in equilibrium with the concentration C_e in the solution phase. Thus q_e has the same units as C_e.

According to the *Freundlich model*, the concentration of the adsorbate in the two phases, i.e. the solution and starch phases, are related by the equation:

$$Q_e = K_F C_e^{1/n} \qquad (6.14)$$

where K_F and n are the model parameters. Note that n has no units and $K_F = (Q_e/C_e^{1/n})$ has the awkward unit [kg$^{(1 - 1/n)}$m$^{3/n}$(kg adsorbate)$^{-1}$]. It may be noted that $(1/n)$ indicates the intensity of adsorption. If $0 < 1/n < 1$, the adsorbate partitions favourably on to the adsorbent, and if $1/n > 1$, then the partition is unfavourable. Freundlich isotherm is merely based on an empirical power law model and does not provide any in-depth insights into the mechanism of the process. It also suggests that as C_e increases Q_e will also increase, if the adsorption is favourable. This is often not the case and experimental results have shown that beyond a certain value of C_e, Q_e hardly changes.

The Langmuir isotherm is closer to experimental data in many cases, and according to this isotherm:

$$Q_e = \frac{Q_M C_e}{\frac{1}{K_L} + C_e} \qquad (6.15)$$

where Q_M [kg (kg adsorbent)$^{-1}$] and K_L (m^3 kg^{-1}) are the model parameters. The similarity in the natures of Eqs. (6.15) and Eq. (5.19) describing the rate of enzyme catalysed reactions is striking. When C_e is small, Q_e varies linearly with C_e, whereas for very high value of C_e, Q_e becomes independent of C_e – which is an experimental observation in a number of cases involving a variety of adsorbates and adsorbents. Moreover Q_M represents the maximum value of Q_e, and $(1/K_L)$ represents the value of C_e when $Q_e = Q_M/2$. These deductions follow the arguments given in Chap. 5 in the paragraph below Eq. (5.19). The Langmuir's constants for a given adsorbate-adsorbent system can be determined by fitting the experimental data Q_e versus C_e to Eq. (6.15) using any software to obtain Q_M and K_L. However, a simple way to do this would be to make a double reciprocal plot, which essentially results by linearising Eq. (6.15) after taking the reciprocals of both sides as follows:

$$\frac{1}{Q_e} = \frac{\frac{1}{K_L} + C_e}{Q_M C_e} = \left(\frac{1}{K_L Q_M} \right) \frac{1}{C_e} + \frac{1}{Q_M} \qquad (6.16)$$

If $(1/Q_e)$ is plotted against $(1/C_e)$ (hence known as the double reciprocal plot), the graph will be linear with slope $= [1/(K_L Q_M)]$ and y-intercept of $1/Q_M$, which allows both Q_M and K_L to be estimated from the experimental data. It would be instructive to undertake a similar linearization of Eq. (5.19) for enzyme kinetics and examine the variables to be plotted in order to obtain a straight line to establish the slope and y-intercept of that line. Incidentally, when the Michaelis-Menten equation is linearised thus, the plot is called Lineweaver-Burk plot.

As explained in Fig. 6.3, the Langmuir isotherm applies when a monolayer of adsorbate is formed on a given adsorbent. It is clearly not beyond the realms of possibility to encounter situations where multiple layers are formed, say, one on top of the other. In such cases, the isotherm does not flatten but continues to progressively increase whilst taking a variety of different shapes. Each of such shapes can command a number of theories and mathematical models to describe it and literature is replete with such theories and more seem to be appearing all the time. One of the theories used to describe multi-layer adsorption, which seems to work for adsorption from solutions, is the Guggenheim-Anderson-de Boer (GAB) model, i.e. Eq. (6.6) which, as stated earlier in Sect. 6.2, is a modification of BET theory (Ebadi et al. 2009). When applied to adsorption, Eq. (6.6) becomes:

$$\frac{Q_e}{Q_M} = \frac{C_{GAB} K_{GAB} C_e}{(1 - K_{GAB} C_e)(1 - K_{GAB} C_e + C_{GAB} K_{GAB} C_e)} \qquad (6.17)$$

Equation (6.17) has three model parameters: C_{GAB}, K_{GAB} and Q_M, which can only be obtained by fitting experimental data on Q_e versus C_e to the above equation.

There are innumerable other equations used to describe isotherms. It is important to note that most of these models have been proposed for gas adsorption on solid surfaces, and their applicability to food systems, which generally involve adsorption of solutes from solutions, is limited. Yet, a number of these models have been used successfully for food system such as the adsorption of polyphenols, antioxidants and vitamin C mainly because of their empirical nature and the number of model

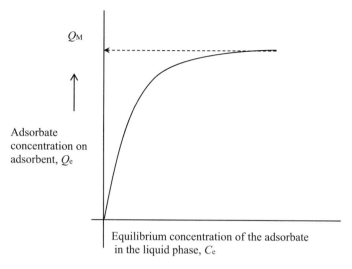

Fig. 6.3 Langmuir isotherm plotting adsorbate concentration on the adsorbent against adsorbate concentration in the solution at equilibrium. As C_e increases, more adsorbate molecules occupy available sites on the adsorbent surface to increase Q_e. After all available adsorption sites are occupied, a monomolecular layer of adsorbate fully covers the adsorbent, and no further increase in Q_e is possible

parameters involved which can be adjusted to fit the experimental data. Caution must be exercised while interpreting the physical significance of the model parameters because their significances need not be the same in gas-solid and liquid-solid systems. A relatively simple isotherm which has been applied to food systems is the *Temkins model*, according to which:

$$Q_e = \frac{RT}{b} \ln\left(K_T C_e\right) \tag{6.18}$$

where b is a model parameter related to the heat of adsorption (J mol^{-1}) and K_T is another isotherm parameter with units inverse of C_e (m^3 kg^{-1})

The *thermodynamics properties of adsorption* can also be deduced from the Langmuir isotherm data. From Eq. (6.1), the Gibbs free energy is $RT \ln K$ where K is the equilibrium constant for the adsorption process. It is necessary to note that K has no units. A number of research papers have used K_L in place of K and this makes the results of such studies questionable. Zhou and Zhou (2014) have re-derived Langmuir isotherm on the basis that the adsorbate molecules substitute solvent molecules which initially occupy the active sites on an adsorbent. Based on this analysis, Zhou and Zhou (2014) have shown that if K_L has the unit of m^3 kg^{-1}, as in Eq. (6.15), the true equilibrium constant K is given by:

$$K = K_L \times M_{adsorbate} \times 55.5 \times 10^6 \tag{6.19}$$

It may be noted that the molar density of water is 55.5 mol L^{-1}. If K_L has the unit of L mg^{-1}, which is common in practice, 10^6 in Eq. (6.19) will be replaced by 10^3.

Using Eq. (6.1), and the definition $\Delta G^0 = \Delta H^0 - T\Delta S^0$ at a temperature T,

$$\ln K = -\frac{\Delta G^0}{RT} = -\frac{\Delta H^0}{RT} + \frac{\Delta S^0}{R} \qquad (6.20)$$

where ΔH^0 and ΔS^0 are the enthalpy and entropy of adsorption, respectively. The enthalpy of adsorption can be experimentally determined if the values of K as a function of T are known. Based on Clausius-Clapeyron equation (Eq. 6.8), ΔH^0 can be obtained from the gradient of the plot of $\ln K$ against $1/T$. Eq. (6.20) can then be used to determine the entropy of adsorption. It is necessary to note that ΔH^0 has been taken to be independent of T in Eq. (6.20) which is quite reasonable. Further, the linearity of the plots of $\ln K$ against $1/T$ has been checked for many food systems over a relatively narrow range of T values. This is because the cooking temperatures invariably range between 60 and 120 °C. Even though this represents a two-fold temperature variation, it only represents a 15% variation on the Kelvin scale which is used in the plot of $\ln K$ against $1/T$. One must therefore be very careful whilst interpreting the slopes and intercepts of such plots. The values of ΔH^0, ΔS^0 and ΔG^0 can vary significantly between various papers published in the literature.

Problem 6.4 **The total organic carbon (TOC) load in waste water from a food factory has to be brought down by adsorption on activated carbon from an initial value of 250 mg L^{-1}. An adsorption study was therefore conducted at 25 °C by adding different amounts of activated carbon to a series of flasks, each containing 200 ml waste water, and the following table gives the final level of TOC (i.e. at equilibrium):**

Activated carbon loading (mg)	Equilibrium TOC (mg L^{-1})
804	4.7
668	7.0
512	9.3
393	16.6
313	32.5
238	62.8

1. **Fit the above data to the Langmuir isotherm and determine the model parameters.**
2. **If a molecule of TOC occupies an area of 0.7 nm², what is the specific area of the activated carbon? Assume that the average molecular weight of TOC is 200, and Avogadro's number is 6.023×10^{23}.**
3. **Estimate the standard Gibbs free energy for adsorption.**
4. **Also fit the data to the Freundlich isotherm and determine the model parameters.**

5. **How much activated carbon is needed for 1 L of waste water if the concentration of TOC is to be reduced from 200 mg L^{-1} to a third of this value?**

Solution It is, first of all, necessary to construct the Q_e versus C_e table. The value of C_e is the second column in the given table. The Q_e value can be obtained by performing a mass balance. To illustrate the calculation, consider the data in the first row. The equilibrium TOC in the water is 4.7 mg L^{-1}. Since the initial concentration is 250 mg L^{-1}, the mass of TOC adsorbed in 200 ml of the solution is: $[250–4.7] \times 200/1000 = 49.06$ mg. Therefore $Q_e = 49.06/804 = 0.06$ mg (mg activated C)$^{-1}$. Thus Q_e values can be filled out for the entire tabulated data as shown below.

C_e	Q_e
mg L^{-1}	mg (mg activated C)$^{-1}$
4.7	0.060
7.0	0.073
9.3	0.094
16.6	0.119
32.5	0.139
62.8	0.157

1. The Langmuir isotherm can be constructed from the Q_e versus C_e data by using the double reciprocal method where $1/Q_e$ is plotted against $1/C_e$ to yield a line. This can be done using a spreadsheet. The slope of the plot $[1/(K_L \cdot Q_M)] = 53.83$ and y-intercept of $1/Q_M = 5.41$; and $R^2 = 0.99$. Thus $K_L = 0.1$ L mg^{-1}, and $Q_M = 0.185$ mg (mg activated C)$^{-1}$.
2. Q_M represents the concentration of a monomolecular layer of TOC $= 0.185$ mg/mg activated C $= [0.185 \times 10^{-3}/200] \times 6.023 \times 10^{23} \times 0.7 \times 10^{-18}$ m^2 per mg activated C $= 0.33$ m^2 mg^{-1} activated C.
3. The true equilibrium constant will be given by Eq. (6.19) with 10^6 replaced by 10^3:

    $K = 0.1 \times 200 \times 55.5 \times 10^3 = 1,110,000$. The Gibbs free energy is: $\Delta G^0 = -RT \ln(1110000) = -8.33(298) \ln (1110000) = -34,554$ J mol^{-1}.
    $1/n = 0.37$. $R^2 = 0.94$.
4. In other words, the Freundlich isotherm is given by: $Q_e = 0.037 \, C_e^{0.37}$. Since the equilibrium concentration is reduced from 200 to a value that is one third, $C_e = 66.67$ mg L^{-1}. Therefore $Q_e = 0.037(66.67)^{0.37} = 0.175$ mg (mg activated C)$^{-1}$. The amount to be adsorbed per litre is $(200–66.67) = 133.33$ mg. Therefore the amount of activated C needed is $133.33/0.175 = 761.85$ mg of activated C.

Problem 6.5 Green tea polyphenols are adsorbed on microparticles of starch and the dimensionless equilibrium constant K (based on the Langmuir isotherm parameter K_L) values, as a function of absolute temperature, are as follows:

Temperature (K)	$K \times 10^{11}$ (dimensionless)
291	1.55
296	1.27
298	1.25
303	1.10
308	0.96

Estimate ΔG^0, ΔH^0 and ΔS^0.

This is a direct application of Eq. (6.20). By plotting ln K against $1/T$, the gradient is $-\Delta H^0/R$ and intercept is $\Delta S^0/R$. By plotting the above graph on a spreadsheet, $-\Delta H^0/R = 2427$ and $\Delta S^0/R = 17.4$. With R $= 8.33$ J mol^{-1} K^{-1}, $\Delta H^0 = -20.2$ kJ mol^{-1}, and $\Delta S^0 = 144.9$ J mol^{-1} K^{-1}.

6.4 Determination of Thermodynamic Equilibrium Parameters from Kinetic Studies

Statistical thermodynamics can be used to deduce thermodynamic parameters from studies on reaction kinetics. The well-known Eyring Equation relates the reaction rate constant with temperature (K), just as the Arrhenius equation does (see Chap. 5, Sect. 5.2.1). But while the latter is empirical, Eyring's model has a statistical thermodynamics basis, and relates the rate constant to the Gibbs free energy of activation, as follows:

$$k = \frac{\kappa k_B T}{h} K^* = \frac{\kappa k_B T}{h} e^{-\frac{\Delta G^*}{RT}} = \frac{\kappa k_B T}{h} e^{-\frac{\Delta H^*}{RT}} e^{\frac{\Delta S^*}{R}} \qquad (6.21)$$

In the above equation, κ is known as the transmission factor which is dimensionless and represents the fraction of the flux through the transition state which proceeds to form the product without re-crossing this stage. The value of κ is invariably assumed to be 1. k_B represents the Boltzmann constant $= 1.38 \times 10^{-23}$ J K^{-1}; $h = 6.63 \times 10^{-34}$ J s is the Plank's constant; and K^* is the equilibrium constant of the activation complex which is dimensionless. In the latter part of Eq. (6.21), K^* has been expressed in terms of the Gibbs free energy of activation by noting that $\Delta G^* = -RT \ln K^*$, and $\Delta G^* = \Delta H^* - T \Delta S^*$. Equation (6.21) can be linearised by taking natural logarithm of both sides, to give:

$$\ln\left(\frac{k}{T}\right) = \left[\ln\left(\frac{k_B}{h}\right) + \frac{\Delta S^*}{R}\right] - \frac{\Delta H^*}{R}\left(\frac{1}{T}\right) \qquad (6.22)$$

Equation (6.22) can be used to determine ΔH^* and ΔS^* by regression of ln (k/T) against $(1/T)$, which shows how reaction kinetics data can be used to determine equilibrium thermodynamic parameters.

Problem 6.6 The entropy and enthalpy of activation of chlorophyll during thermal processing of ground mint leaves is reported to be $-0.22\,\text{kJ mol}^{-1}\,\text{K}^{-1}$ and $20.9\,\text{kJ mol}^{-1}$, respectively (data taken from Gaur et al. 2007). Assuming Boltzmann constant $= 1.38 \times 10^{-23}\,\text{J K}^{-1}$, Plank's constant $h = 6.63 \times 10^{-34}\,\text{J s}$ and the ideal gas constant to be $8.34\,\text{J mol}^{-1}\,\text{K}^{-1}$ estimate what fraction of chlorophyll will get deactivated if ground mint leaves are heated at $80\,^\circ\text{C}$ for 2 mins.

Solution   According to Eq. 6.21, the rate constant is given by:

$$k = \frac{k_B T}{h} e^{-\left(\frac{\Delta H^*}{RT}\right)} e^{\left(\frac{\Delta S^*}{R}\right)}$$

With $\Delta H^* = 20,900\,\text{J mol}^{-1}$, $\Delta S^* = -220\,\text{J mol}^{-1}$, $k_B = 1.38 \times 10^{-23}\,\text{J K}^{-1}$, $h = 6.63 \times 10^{-34}\,\text{J s}$, $T = 353\,\text{K}$ and $R = 8.34\,\text{J mol}^{-1}\,\text{K}^{-1}$, the rate constant works out to $k = 0.021\,\text{s}^{-1}$. Assuming first order degradation of chlorophyll, the fraction of chlorophyll remaining after heating for $t = 120\,\text{s}$ is $e^{-kt} = 0.08$. Therefore fraction deactivated is $1{-}0.08 = 0.92$ or 92% of chlorophyll will get deactivated.

6.5   Entropy-Enthalpy Compensation Effects

Unlike normal chemical reactions, the biochemical reactions occurring during food processing cannot, in general, be represented by simple balanced chemical equations. Chlorophyll degradation occurring during heating of leaves such as spinach and mint, or protein denaturation caused by heat are a typical examples. For a start, chlorophyll or proteins in a food do not have simple molecular formulae, and the exact degradation products are also unclear. However, the thermal behaviour of chlorophyll or proteins under different conditions (e.g. pH conditions or in different solvents) represent a series of closely related chemical reactions, which exhibit a correlation – often a linear relationship – between the enthalpy and entropy changes, known as *enthalpy-entropy compensation*. In general, if ΔH_i represents the enthalpy change of any one of a series of reactions and ΔS_i the corresponding entropy change, then:

$$\Delta H_i = \alpha + \beta \Delta S_i \tag{6.23}$$

It is evident that if such a relationship exists for a series of different reactions, i.e. all the reactions possess the same values of α and β, this linear relationship helps to estimate thermodynamic property changes *a priori*. Further, this relationship also suggests that the mechanisms of the reactions are also similar. Moreover, α has the unit of enthalpy (J mol^{-1}), ΔS_i has the unit $(\text{J mol}^{-1}\,\text{K}^{-1})$ and β has the unit of temperature (K). β is also known as the *compensation temperature*.

Example 6.1 The enthalpy-entropy compensation equation for the binding of zinc on specific amino acids was experimentally determined in a salt solution at 25 °C to be: $\Delta H = 292\Delta S - 21253$, where the enthalpy is expressed in J mol^{-1} and the entropy in J mol^{-1} K^{-1}. If the binding of zinc on peptides and whey proteins follow the same enthalpy-entropy compensation, estimate the standard Gibbs free energy change for the peptide glutathione and β-lactoglobulin if the enthalpy changes are determined to be -16.9 and 0.8 kJ mol^{-1}, respectively. Also estimate the equilibrium constants for zinc binding on these materials.

Solution It may be noted that zinc binding on glutathione is exothermic while that on β-lactoglobulin is marginally endothermic. Using the compensation equation, the entropy changes (ΔS_i) for glutathione and β-lactoglobulin are 14.9 and 75.5 J mol^{-1}. The Gibbs free energy change is therefore given by $\Delta G_i = \Delta H_i - T\Delta S_i$. With $T = 298°$K, the Gibbs free energy changes are $-21,340$ and $-21,699$ J mol^{-1} respectively. If the binding of the zinc to a ligand L$^-$ is represented as: $Zn^{2+} + L^- \Leftrightarrow ZnL^+$, the equilibrium constants for the binding, K, can be defined as $K = \frac{[ZnL^+]}{[Zn^{2+}][L^-]}$. Since $\Delta G^0 = -RT \ln K$, the equilibrium constant values for the binding of zinc on glutathione and β-lactoglobulin are 5.36×10^4 and 6.19×10^3 L mol^{-1}, respectively, assuming that all concentrations in the equilibrium constant are represented as molarity, which is the norm.

References

Bell LN, Hageman MJ (1994) Differentiating between the effects of water activity and glass transition dependent mobility on a solid state chemical reaction: aspartame degradation. J Agric Food Chem 42:2398–2401

Caurie M (2007) Hysteresis phenomenon in foods. Int J Food Sci Technol 42:45–49

Condon JB (2006) Theories behind the chi plot. In: Condon JB (ed) Surface area and porosity determinations by physisorption. Elsevier, Oxford, pp 91–125

Ebadi A, Soltan JS, Khudiev MA (2009) What is the correct form of BET isotherm for modeling liquid phase adsorption? Adsorption 15:65–73. https://doi.org/10.1007/s10450-009-9151-3

Gaur S, Shivhare US, Sarkar BC, Ahmed J (2007) Thermal chlorophyll degradation kinetics of mint leaves puree. Int J Food Prop 10:853–865

Pollio ML, Kitic D, Resnik SL (1996) Aw values of six saturated salt solutions at 25°C. Re-examination for the purpose of maintaining a constant relative humidity in water sorption measurements. Lebensm-Wiss u-Technol 29:376–378

Zhao X, Zhao X (2014) The unit problem in the thermodynamic calculation of adsorption using the Langmuir equation. Chem Eng Commun 201:1459–1467. https://doi.org/10.1080/00986445.2013.818541

Further Reading

http://library.metergroup.com/Application%20Notes/13947_Fundamentals%20of%20Moisture%20Sorption%20Isotherms_Web.pdf

Reichelt S (ed) (2015) Affinity chromatography. Springer, New York/Heidelberg

Smith JM, Van Ness HC, Abbott MM, Swihart MT (2017) Chemical engineering thermodynamics, 8th edn. McGraw Hill Education, New York

Chapter 7
Thermal Processing of Foods

Aim The aim of this chapter is to illustrate the principles underpinning thermal processing of foods, in particular, focusing on how thermal processing conditions determine the reduction in microbial load, which in turn ensures food safety and extends its keeping quality. The chapter addresses in-container processing of foods, as well as continuous processing in tubular configurations followed by filling of packages in a contained environment (also commonly known as aseptic processing).

7.1 Introduction

Thermal processing may have a variety of objectives, but reducing the microbial load in a food is a key objective. This reduction may target specific microbial species, but foods invariably contain mixed microbial populations and thermal processing results in lowering the total microbial count, with the extent of reduction depending on the intensity of the thermal process employed. Thermal processing of foods is inevitably accompanied by other physicochemical changes, such as cooking, texture softening, changes in nutrient profiles, changes in flavour and inactivation of enzymes. Some of the changes induced by thermal processing may well be desirable and other may not be so. For instance, if *blanching* or *pasteurisation* is the objective of thermal processing, then enzyme inactivation is a key purpose of thermal processing, because its active presence can accelerate food spoilage and reduce its shelf life. Of course, microbial load is also lowered during blanching and pasteurisation, with pathogens and other heat sensitive microorganisms getting destroyed in these processes. It is necessary to note that blanching and pasteurisation are relatively milder forms of thermal processing with the temperatures adopted being as low as 70–90 °C. If the goal of thermal processing is to achieve *commercial sterility*, i.e. eliminate virtually all microorganisms present in

© Springer Nature Switzerland AG 2022
K. Niranjan, *Engineering Principles for Food Process and Product Realization*,
Food Engineering Series, https://doi.org/10.1007/978-3-031-07570-4_7

the food, then the processing temperatures employed are in the range 120–140 °C. Clearly, such high temperatures can only be achieved without causing significant phase changes by operating processes under pressure. Moreover, the processing times are also very short at around 6–10 s. In general, thermal processing times are lower at higher temperatures, which is essential to retain the nutrient and flavour profiles of foods. As mentioned above, blanching and pasteurisation are relatively milder forms of thermal processing whereas *canning*, *baking*, *roasting* and *frying* are more severe processes.

7.1.1 Blanching

The primary purpose of blanching is to inactivate enzymes in fresh fruit and vegetable produce. It is generally not intended to be the sole purpose of preservation, but used a pre-treatment step prior to freeing, dehydration and other processes, which, on their own are unable to completely inactivate food quality deteriorating enzymes such as lipoxygenase, polyphenol oxidase, polygalacturonase and chlorophyllase. In addition to enzyme inactivation, blanching also achieves: (a) a reduction in surface microbial contamination, (b) softening of fruit and vegetable tissues to facilitate filling into containers, and (c) removal of air from intercellular spaces which is critical in canning process.

Blanching is generally carried out at temperatures up to 100 °C by using hot water or steam as the blanching medium. Some processes also employ microwaves to blanch. Hot water is commonly used for blanching especially if salt, sugar or texture modifying agents such as calcium chloride are required in the blanching medium, but this process has a major drawback caused by solutes leaching out of the fruit or vegetable. Steam blanching is widely used in practice and in this method the food is carried through steam on a wire mesh conveyer belt or a rotary cylinder, with the belt speed or rotational speed controlling the steam exposure time. Although steam heating is more efficient than hot water to blanch a kilogram of food, this process often results in non-uniform heating with the food surface being significantly over heated in comparison to the centre. A modification of this process is called *individual quick blanching* where blanching occurs in two stages: an initial stage where a single layer of food is heated to inactivate the enzyme, followed by a deep bed of the food being held in heat in order for the temperature at the centre of the food to attain the enzyme inactivation temperature. The *peroxidase test* is commonly used to ascertain the efficacy of blanching because this is a relatively heat resistant enzyme and its absence can confirm enzyme inactivation. The test consists of adding hydrogen peroxide and guaiacol solutions to a sample of the food, with the development of brown colouration indicating peroxidase activity.

7.1.2 Pasteurisation

Like blanching, pasteurisation is also a relatively mild form of thermal processing used in the case of foods that are predominantly in the liquid state, e.g. milk, fruit juices and beverages. Pasteurisation destroys enzymes, and relatively heat sensitive micro-organisms which include pathogens, yeast and moulds – thus extending the keeping quality of such foods. If milk has to be pasteurised, typically the temperatures employed are 72 °C for 15 s (the process being known as High Temperature Short Time process or HTST) or 63 °C for 30 min (Low Temperature Long Time process or LTLT). If higher temperatures are used, the processing time required to achieve pasteurisation is reduced sharply. For instance, if the temperature employed is 90 °C, then a process time of 0.5 s should suffice. Pasteurisation cannot destroy spore forming microorganisms as well as *thermoduric* microorganisms (i.e. microorganisms which survive high temperatures but do not grow at these temperatures) and *thermophilic* microorganisms (i.e. microorganisms capable of surviving as well as growing at high temperatures). Pasteurisation can be undertaken in a batch mode or continuous mode.

7.1.3 Sterilisation

Normally, there is no requirement for foods to be perfectly sterile. Most practical processes aim to achieve what is known as *commercial sterility* – heat processing designed to kill substantially all microorganisms and spores, which if present, would be capable of growing in the food under normal storage conditions. Thus, commercial sterility implies less than absolute destruction of all microorganisms and spores, but any remaining would be incapable of growth in the food under existing conditions. For the purpose of achieving commercial sterility, *Clostridium botulinum* is considered to be the target organism because of its heat resistant spore forming nature. If a process is capable of destroying this species substantially, it would be safe to assume that all other microorganisms are also destroyed. Sterilisation processes are inherently severe and also bring about changes in nutritive and sensory properties.

Typically, two types of sterilised products are commercially available. (1) *In-container* processed products, where the product is first packaged in containers such as cans and bottles and subsequently heated by steam in retorts for a stipulated period of time, followed by cooling. Canned foods are produced thus. (2) *Aseptically* processed products, which involve thermal processing of the product and packaging separately, followed normally by aseptic filling of the packages. *Ultra High Temperature (UHT)* processed products are made this way. Typically, in UHT processing the food is heated to temperatures around 140 °C for 2–5 s. In contrast, canning is a much longer process and at temperatures of 120 °C, the process times could be 20 min or greater depending on the type of food. The principles which determine the time-temperature combinations employed in practice depend on microbial and nutrient inactivation kinetics, which will be considered in the following sections.

7.2 Kinetics of Microbial Inactivation

Generally, microbial inactivation is considered to be a *first order process* (see Chap. 5, Sect. 5.2) according to which the rate at which microorganisms are thermally inactivated at a given temperature, is proportional to their number. Mathematically, if N represents the microbial population at any time, and N_0 the number present initially,

$$-\frac{dN}{dt} = -kN \text{ whence } \ln\left(\frac{N_0}{N}\right) = kt \text{ or } N = N_0 e^{-kt} \tag{7.1}$$

Alternatively, the microbial population N at any time t can also be expressed in terms of decimal logarithm instead of natural logarithm:

$$2.303 \log\left(\frac{N_0}{N}\right) = kt \tag{7.2}$$

The first order rate constant k (s^{-1} or min^{-1}) is essentially the constant of proportionality in the first order differential equation given in Eq. (7.1). While this constant of proportionality seems somewhat abstract, it is customary to express it in terms of a more physically imaginable parameter. This is done by determining the time taken, D, for the microbial population to fall from N_0 to a tenth of its initial value (i.e., $N = N_0/10$). Thus, by taking $N_0/N = 10$ in Eq. (7.2)

$$D = \frac{2.303}{k} \tag{7.3}$$

D is known as the *decimal reduction time* and it represents the time taken for the microbial numbers to *fall to a tenth* or *by a log cycle*. In other words, if a food is maintained at a given high temperature for a time D, the microbial numbers in it will fall by 90%. Equation (7.2) can also be expressed in terms of D as follows:

$$D \log\left(\frac{N_0}{N}\right) = t \tag{7.4}$$

D is also is a characteristic time for a microbial species at a given temperature, and typical values at 121 °C (or 250 °F) are reported in Table 7.1. According to Eq. (7.2), a plot of log N against time is linear; see Fig. 7.1.

It is also clear from Fig. 7.1 that the decimal reduction time depends on the temperature and one expects it to drop with increasing temperature because the

Table 7.1 Typical values of the decimal reduction time D for different species at 121 °C

Bacterial groups	Approx. heat res.	Reference
Low-acid and semi-acid foods (pH > 4.5)	D_{121} **(min)**	
Thermophiles		
B. stearothermophilus	4.0–5.0	Hodges (2018)
C. thermosaccharolyticum	3.0–4.0	Naczk and Artyukhova (2020)
C. nigrificans	2.0–3.0	Setlow and Johnson (2019)
Mesophiles		
C. botulinum	0.10–0.20	Gonzalez-Fandos and Laorden (2019)
C. sporogenes	0.10–1.5	Soni et al. (2020)
Acid foods (pH 4.0–4.5)		
Thermophiles		
B. coagulans	0.01–0.07	Rayman Ergün and Baysal (2019)
Mesophiles	D_{100} **(min)**	
B. polymyxa and B. macerans	0.10–0.50	Barroso et al. (2020); Zahirinejad et al. (2021)
C. pasteaurianum	0.10–0.50	Gabriel-Barajas et al. (2022)
High-acid foods (pH < 4.0)	D_{65} **(min)**	
Mesophilic non-spore-bearing bacteria		
Lactobacillus spp; *Leuconostoc* spp., and yeast and moulds	0.50–1.00	Abdel Gawad et al. (2021); Azizi (2018); Montanari et al. (2019); Silva (2019)

Adapted from Hartel and Heldman (1997)

Fig. 7.1 Variation of microbial numbers with time indicating a linear relationship between log N and time t with the gradient being equal to $-1/D$. An increase in temperature results in lower values of D and steeper gradients

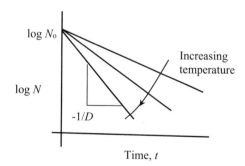

microorganisms would perish faster at higher temperatures. The temperature rise required for the D value to reduce to one tenth of its original value is known as the Z value or *thermal resistance constant* of a microbial species. Thus Z value, which has temperature units (°C), indicates the resistance of the microorganism to a change in temperature, and a higher Z value corresponds to slower response to temperature changes (i.e. greater heat resistance among microbes). If the decimal reduction time is known at a temperature T_1 and its value at a temperature T_2 is to be found out, then:

$$D_{T_2} = D_{T_1} 10^{\frac{T_1-T_2}{z}} \tag{7.5}$$

In a practical thermal process, there is always a stated reduction in microbial population which may refer to pathogens or spoilage bacteria or any other targeted bacteria levels. The time taken for the stated reduction in microbial numbers is known as the F value, which effectively represents the thermal treatment time at any given temperature. F values are normally expressed as multiples of D values. For example, a $12D$ or 12 log cycle reduction is most commonly used for low acid food (pH > 4.6), and it can be interpreted as the time taken to reduce the bacterial count from N_0 to $10^{-12} N_0$. It can also be interpreted as reducing the probability of spoilage due to microbial activity to 1 in $(10^{12}/N_0)$. So, if 10 million microorganisms are present initially, there is a 1 in 10^5 chance of spoilage. In other words, if this is a canning process, the probability of a can being at risk is one in 10^5.

For acidic foods (pH < 4.6) it is common practice to use $F = 5D$ process, which is the time/temperature process that will reduce the targeted species like *Bacillus stearothermophilus* population by 5 log cycles. In other words, the time required to reduce *Bacillus* count from N_0 to $10^{-5} N_0$. A $5D$ process also implies heating the food for significantly shorter times than $12D$ process which will also help retain the natural attributes of the food if it is so desired.

Since F is a multiple of D for a desired level of microbial inactivation at a given temperature T, based on Eq. (7.5), F values at two temperatures for equivalent levels of microbial inactivation are given by:

$$F_{T_2} = F_{T_1} 10^{\frac{T_1-T_2}{z}} \tag{7.6}$$

Even though the holding temperature is fixed, a thermal process consists of heating, holding and cooling stages, and the net heat treatment as well as the resulting microbial inactivation is a result of all three stages. The *integrated lethality value*, F_0, is defined as the time, at a constant temperature T_{ref}, that produces the same microbial inactivation effect as the actual thermal process. Ball and Olson (1957) showed that F_0 is given by:

$$F_0 = \int_0^t 10^{\frac{T-T_{\text{ref}}}{z}} dt \tag{7.7}$$

Comparison of the calculated F_0 values of a process with the time required to destroy a given percentage of the target microbial population (i.e. the target F_0 value), is the basis for the thermal process design. For example, the F_0 target value for the *Clostridium botulinum* spore in a sterilization process at a temperature of 121.1 °C (250 °F) is at least 6 min An illustration of typical temperature profiles during retorting is shown in Fig. 7.2.

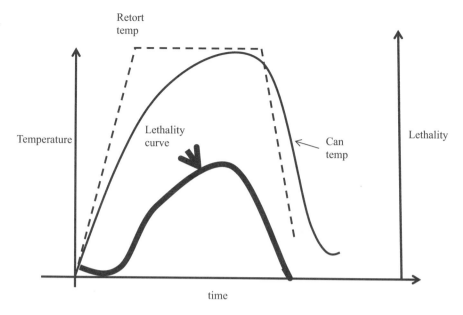

Fig. 7.2 Typical temperature profiles during retorting of cans. The lethality curve is a plot of $\frac{T-T_{ref}}{Z}$ against t, and F_0 is the area under the curve

Example 7.1 A microorganism has a D_{80} value (i.e. the decimal reduction time at 80 °C) of 1.5 min and a Z value of 8 °C. Calculate its D value at 90 °C. A sample containing 10^{12} ml^{-1} of this micro-organism is heated for 30 s at 90 °C. Estimate the final population in the heated product.

Solution: This problem is based on the application of Eq. (7.5). Thus, $D_{90} = D_{80}10^{(80 - 90)/Z}$, with $Z = 8$ °C and $D_{80} = 1.5$ min, we have $D_{90} = 0.084$ min. (Note that the decimal reduction time is lower at higher temperature). The thermal death time $F = 30$ s $= 0.5$ min; this must be equal to: $D_{90} \log_{10}(N_0/N)$ according to Eq. (7.4). With $N_0 = 10^{12}$ ml^{-1}, we have: $0.5 = 0.084 \log_{10}\frac{10^{12}}{N}$ giving $N = 1.11 \times 10^6$ ml^{-1}.

Example 7.2 Based on Eq. (7.6), the F_T value (expressed in minutes) delivered by a process at a temperature T °C is related to F_{ref} (the reference value) at $T_{ref} = 121$ °C for the same level of microbial inactivation, by the relationship: $F_T = F_{T_{ref}}10^{\frac{T_{ref}-T}{Z}}$. **If $F_{100} = 450$ min and $F_{150} = 0.0078$ min are equivalent processes (i.e. delivering the same levels of microbial inactivation) at 100 °C and 150 °C, respectively, estimate (i) the z value for the microorganism targeted (ii) F_{ref} value.**

Solution: From the data, $F_{100} = F_{T_{ref}} 10^{\frac{121-100}{Z}} = 450$, and $F_{150} = F_{T_{ref}} 10^{\frac{121-150}{Z}} = 0.0078$. Dividing the F values at the two temperatures, we eliminate $F_{T_{ref}}$ and obtain $10^{50/Z} = 57692.3$, whence $Z = 10.5$ °C. Using this value of Z in any one of the two equations, we have $F_{T_{ref}} = 4.51$ min.

Example 7.3   The microbial load is 60 spores per can prior to sterilisation. If the decimal reduction time of the spores at 121 °C is 1.5 min, and the acceptable level of safety risk is 1 can in 100,000, what should be the F value of the thermal process, i.e. how long must the cans be heated in a retort at 121 °C?

Solution: The acceptable probability of risk is 1 in 100,000 i.e. 10^{-5}. We know that an xD process gives a risk probability of 1 in $10^x/N_0 = 1/(10^x/60) = 10^{-5}$. Solving for x, we have $x = 6.8$. Therefore $F = 10.2$ min.

Example 7.4   The following time-temperature history was noted during the pasteurisation of a citrus fruit juice. Estimate the equivalent F value at 95 °C, assuming the Z value to be 10 °C.

Time (s)	Temperature (°C)
0	20
5	48
8	63
12	75
13	82
16	90
25	90
30	79
34	60
40	48
50	33

Solution: From Eq. (7.7), $F_0 = \int_0^t 10^{\frac{T-T_{ref}}{Z}} dt = \int_0^t 10^{\frac{T-95}{10}}$. The integral can be approximated to $\sum_0^t 10^{\frac{T-95}{10}} \Delta t$ and evaluated to obtain $F_0 \cong 3.59$ s.

Example 7.5   What is the F_0 value for the sterilisation of spinach puree in a can, given the following time-temperature profile (Yanniotis 2007). Assume the reference temperature to be 121.1 °C and Z value for the targeted microorganism to be 10 °C. If $D_{121.1}$ for the targeted microorganism is 0.2 min, comment on the adequacy of the thermal process.

Time (min)	Temperature (°C)
0	50
3	80

(continued)

Time (min)	Temperature (°C)
5	100
8	115
11	119
13	121
14	119
15	110
17	85
19	60
20	50

Solution: The methodology adopted is the same as Example 7.4, $F_0 =$
$\int_0^t 10^{\frac{T-T_{ref}}{Z}} dt = \int_0^t 10^{\frac{T-121.1}{10}}$. The integral can be approximated to $\sum_0^t 10^{\frac{T-121.1}{10}} \Delta t$ and evaluated to obtain $F_0 \cong 5.25$ min. Since $D = 0.2$ min, the process time as a multiple of D is $5.25/0.2 = 26.25$. A $26.25D$ process is far more severe than the normally used $12D$ process standard. This will be detrimental to product quality as Example 7.6, below, demonstrates.

7.3   Kinetics of Nutrient Inactivation

Along with microbial inactivation, an inevitable effect of thermal processing is nutrient inactivation. Thus, key food components like vitamins, chlorophyll and antioxidants also end up getting inactivated. The approach to modelling nutrient inactivation is very similar to microbial inactivation. In other words, even nutrient inactivation is assumed to follow first order kinetics and is described by an equation similar to Eq. (7.1) with N being replaced by the nutrient concentration C:

$$-\frac{dC}{dt} = -kC \quad \text{whence} \quad \ln\left(\frac{C_0}{C}\right) = kt \quad \text{or} \quad C = C_0 e^{-kt} \tag{7.8}$$

Likewise, a decimal reduction time D and a thermal resistance constant Z can also be defined for nutrient inactivation, with D representing the time taken for the nutrient concentration to drop to one tenth of its initial value (or time taken for 90% of the nutrient to get inactivated) and Z representing the temperature rise that would result in D value itself becoming one tenth of its starting value. Table 7.2 lists the D and Z values for nutrient inactivation. It is clear from the table that D values for microbial inactivation are significantly lower than the D values for nutrient inactivation, which suggests that microbial inactivation levels are far greater than nutrient inactivation level for a given thermal processing time-temperature combination. It is indeed this

Table 7.2 Effect of thermal processing on nutrient inactivation

Component	Substrate	pH	D_{121} (min)	Z_C (°C or K)	Reference
C. botulinum	–	>4.5	0.10–0.20	8.2–9.1	Rodríguez-Ramos et al. (2021)
B. stearothermophilus	–	>4.5	4.0–5.0	7.0–12.0	Pereira et al. (2019)
Ascorbic acid	Canned peas	–	921	17.8	Hashemi et al. (2019)
Thiamine	Beef puree	–	254	25.4	Amsasekar et al. (2022)
Thiamine	Pea puree	–	247	25.2	Avilés-Gaxiola et al. (2018)
Folates	Apple juice	3.4	492	35.2	Zhao et al. (2021)
Chlorophyll	Peas, blanched	–	13.4	43.5	Zhang et al. (2021)
Chlorophyll	Spinach	Natural	13.0	25.6	Manzoor et al. (2021)
Non-enzymatic apple juice	Natural	384	35.3	Browning	Zhu et al. (2022)
Vitamin B6	Cauliflower	–	411	50.7	Balasubramanian et al. (2019)
Vitamin A	Beef liver puree	–	43.5	26.0	Zhao et al. (2018)
Anthocyanins	Concord grape	3.4	123	53.4	Li and Padilla-Zakour (2021)

Adapted from Hartel and Heldman (1997)
In order to enable comparison, the first two rows give the relevant data for microbial inactivation

fact which makes thermal processing of foods viable, because thermal processing time employed to inactivate substantially all microorganisms present, does not result in significant nutrient inactivation.

Even in the case of nutrient and quality changes, it must be recognised that these changes do not occur exclusively in the holding stage, but these changes can also occur in the heating and cooling stages. Similar to lethality value, the degree of heat treatment with respect to the "quality" factors can be represented by, what is commonly known as the cooking value or *C*-value defined, very similar to Eq. (7.7), as follows:

$$C = \int_0^t 10^{\frac{T-T_{ref}}{Z_C}} \, dt \qquad (7.9)$$

It may be noted that Z_C is the temperature interval over which the decimal reduction in the cooking parameter changes by a tenth. Intuitively, we have assumed that the cooking parameter also follows first order kinetics – this is an oversimplification! It is also possible to find a relationship between Z_C and the *activation energy E* for nutrient inactivation (see Chap. 5, Sect. 5.2.1, Eq. 5.9). If D_{T_1} is the decimal reduction time for nutrient inactivation at temperature T_1, it will be a tenth of this value at temperature $(T_1 + Z_C)$; note that the temperature T_1 is taken in K, whereas Z_C being a temperature difference is the same in K or °C. Since the ratio of the two decimal reduction times is also the inverse ratio of the rate constants k_1 and k_2 according to Eq. (7.3)

$$\frac{D_{(T_1+Z_C)}}{D_{T_1}} = \frac{k_{T_1}}{k_{T_1+Z_C}} = \frac{1}{10} \tag{7.10}$$

According to Arrhenius equation (Eq. 5.9), the rate constant at any given temperature $k_T = A \exp(-E/RT)$, Eq. (7.10) can be written as:

$$\frac{k_{T_1+Z_C}}{k_{T_1}} = e^{\frac{E}{R}\left(\frac{1}{T_1} - \frac{1}{T_1+Z_C}\right)} = 10 \tag{7.11}$$

Taking the natural logarithm of the second part of Eq. (7.11) and simplifying, we have:

$$Z_C = \frac{2.303R}{E} T_1(T_1 + Z_C) \tag{7.12}$$

Equation (7.12) relates the Z value for nutrient destruction with the activation energy, so if one of these parameters is known, the other can be determined. Normally, the process temperatures are significantly over 300 K and Z_C values are significantly lower (see Table 7.2); so no significant errors are introduced if $(T_1 + Z_C) \cong T_1$. Therefore:

$$Z_C = \frac{2.303R}{E} T_1^2 \tag{7.13}$$

Example 7.6 The activation energy for vitamin B_6 destruction in cauliflower is 66,638 kJ kmol^{-1} around normal sterilisation temperature of 120 °C. What is Z_C, given the ideal gas constant is 8.31 kJ kmol^{-1} K^{-1}?

Solution: In Eq. (7.12), $T_1 = 393$ K, $E = 66,638$ kJ kmol^{-1} and $R = 8.31$ kJ kmol^{-1} K^{-1}. Substituting these values $Z_C = 50$ K. If instead of Eq. (7.12), the calculation was based on Eq. (7.13), the approximate value of $Z_C = 44.36$ K.

Example 7.7 If D and Z values for the thermal destruction of chlorophyll in spinach are 13 min at 121.1 °C and 26 °C, respectively, estimate the percentage loss of cholorophyll for the time temperature data given in Example 7.5.

Solution: The C value is given by Eq. (7.9): $C = \int\limits_{0}^{t} 10^{\frac{T-T_{ref}}{Z_c}} dt = \int\limits_{0}^{t} 10^{\frac{T-121.1}{26}} dt$. The

integral can be approximated to $\sum\limits_{0}^{t} 10^{\frac{T-121.1}{26}} \Delta t$ and evaluated to obtain $C = 7.9$ min.

Thus, for the purpose of chlorophyll destruction, the time temperature profile represents heating the can at 121.1 °C for 7.9 min. Since D value for chlorophyll destruction at 121.1 °C is 13 min, $k = 2.303/D = 0.18$ min^{-1}. Therefore the fraction of chlorophyll remaining is: $e^{-kt} = \exp(-0.18 \times 13) = 0.24$. Thus 76% of the chlorophyll is lost in the process, which does not bear well for the time-temperature profile employed in the process. As seen in Example 7.5, the process is very severe even from a microbial sterilisation point of view. The heating times must be shortened.

7.4   Aseptic Processing

As mentioned earlier, in this method, the product is processed outside the packaging. The product is then filled into the packaging aseptically. It is also normally conducted in a continuous manner. The thermal processing can either involve pasteurisation or sterilisation of the product (the latter being known as UHT processing). UHT process requires sterile equipment, sterile product, sterile packaging and sterile environment (Fig. 7.3).

A typical aseptic thermal processing line consists of a heat exchanger to heat the product to the desired temperature. The product is then held at this temperature in the *holding tube* for microbial inactivation to occur by pumping it at such a rate that it remains in the holding tube for a time period which corresponds to the F value for the treatment. If Q m^{3}s^{-1} is the volumetric flow rate of a liquid food through a holding tube of diameter d_h and length L_h, the mean superficial velocity of flow is $u_{mean} = Q/[(\pi/4)d_h^2]$ and the mean residence time in the holding tube is (L_h/u_{mean}). It is well

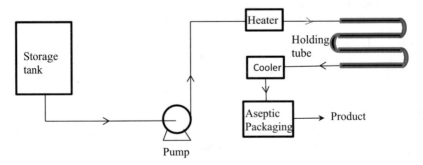

Fig. 7.3 A schematic diagram of an aseptic thermal processing line showing the three key stages of the product being heated, held and cooled, Thermal processing is then followed by aseptic packaging

known from basic fluid mechanics that the fluid elements closer to the axis of the holding tube are moving much faster than the elements moving closer to the tube wall; see Chap. 2, Sect. 2.4. Thus a flow velocity distribution prevails in the holding tube, the average value of which is u_{mean}. The design of the thermal process must take into account the velocity distribution, and must ensure that the fastest moving particles (which spend the shortest period of time in the holding tube) spend sufficient time in the holding tube and achieve the desired microbial inactivation level. It is customary to assume that the maximum flow velocity of the fluid elements flowing along the axis of the holding tube is *twice* the mean velocity, which in turn implies that the minimum residence time in the holding tube is $(L_h/2u_{mean})$. Therefore, it is customary in process design to estimate F as:

$$F = \frac{L_h}{2u_{mean}} \tag{7.14}$$

If the fastest moving elements of the liquid have received the required thermal treatment, it would be safe to assume that all other elements have also spent a time greater than F in the holding tube, which ensures that the product is safe as a whole.

Equation (7.14) assumes that the fluid elements in the holding tube are all at the same temperature, which corresponds to the microbial inactivation temperature of T. However, just as in the case of retorting, microbial inactivation does not merely occur in the holding tube, but also during heating and cooling. Further, the liquid residence time distributions in the heat exchangers as well as in the holding tube induce thermal residence time distributions and an equivalent lethality must be estimated similar to Eq. (7.6).

Estimation of lethality and cooking values can become quite complicated in the case of liquid foods containing suspended particulates. A detailed analysis of this topic has been presented by Ibrahim et al. (2019).

Example 7.8 In the UHT processing at 120 °C, milk is flowing continuously at the rate of 200 L min^{-1} through a holding tube to achieve a 12D reduction in the number of microorganisms. The holding tube is a well-insulated stainless steel pipe, 55 mm internal diameter and 30 m long. If the first order microbial inactivation rate constant is 1.84 s^{-1}, would you expect the holding tube to provide sufficient holding time? The average rate constant for vitamin destruction in milk at 120 °C may be assumed to be 0.001 min^{-1}. How much vitamin would be destroyed in the 12D process?

Solution: The volumetric flow rate of milk $= 200$ L min$^{-1} = 3.33 \times 10^{-3}$ m^{3}s^{-1}. The mean flow velocity through the 55 mm diameter holding tube is: $3.33 \times 10^{-3}/[(\pi/4)(0.055)^2] = 1.4$ m s^{-1}. The maximum flow velocity can therefore be estimated as twice the mean, i.e. 2.8 m s^{-1}. Since the length of the tube is 30 m, the minimum residence time $= 30/2.8 = 10.7$ s. The time for microbial inactivation is 12D. With the rate constant for microbial inactivation being 1.84 s^{-1}, the decimal reduction time D is $2.303/1.84 = 1.25$ s (Eq. 7.3). The process time required for 12D inactivation is 15.02 s. Since the process time is longer than the minimum residence time in the holding tube, the length of the holding tube will not be sufficient. This can be remedied by having a longer tube with length $= 2.8 \times 15.02 = 42.06$ m.

Vitamin remaining in the milk is given by $C/C_0 = e^{-kt}$ where k is the rate constant $= 0.001$ min^{-1} or 1.67×10^{-5} s^{-1}, and $t = 15.02$ s. Substituting these values it is clear that 99.97% of vitamin will be retained and the amount destroyed will be negligible.

References

Abdel Gawad DO, Emara MM, Kassem GM, Mohamed MA (2021) Controlled bio-fermentation by lactobacillus and Lactococcus probiotics for improving quality and safety of Fessiekh (fermented grey mullet). J Aquat Food Prod Technol 31:128–139. https://doi.org/10.1080/10498850.2021.2021340

Amsasekar A, Mor RS, Kishore A, Singh A, Sid S (2022) Impact of high pressure processing on microbiological, nutritional and sensory properties of food: a review. Nutrition & Food Science

Avilés-Gaxiola S, Chuck-Hernández C, Serna Saldivar SO (2018) Inactivation methods of trypsin inhibitor in legumes: a review. J Food Sci 83(1):17–29

Azizi A (2018) Thermal processing determination time for fermented and acidified indigenes Iranian vegetables. J Food Bioprocess Eng 1(2):97–102

Balasubramanian S, Kalne AA, Khan KA (2019) Effects of processing on vitamins in fruits and vegetables. In: *Processing of fruits and vegetables*. Apple Academic Press, pp 297–316

Ball CO, Olson FCW (1957) Sterilization in food technology. McGraw-Hill, New York

Barroso JRM, Mariano D, Dias SR, Rocha RE, Santos LH, Nagem RA, de Melo-Minardi RC (2020) Proteus: an algorithm for proposing stabilizing mutation pairs based on interactions observed in known protein 3D structures. BMC Bioinformatics 21(1):1–21. https://doi.org/10.1186/s12859-020-03575-6

Gabriel-Barajas JE, Arreola-Vargas J, Toledo-Cervantes A, Méndez-Acosta HO, Rivera-González JC, Snell-Castro R (2022) Prokaryotic population dynamics and interactions in an AnSBBR using tequila vinasses as substrate in co-digestion with acid hydrolysates of Agave tequilana var. azul bagasse for hydrogen production. J Appl Microbiol 132(1):413–428. https://doi.org/10.1111/jam.15196

Gonzalez-Fandos E, Laorden AM (2019) Sous vide technology. In: Innovative technologies in seafood processing. CRC Press, pp 263–278

Hartel RW, Heldman DR (1997) Principles of food processing. Springer

Hashemi SMB, Roohi R, Mahmoudi MR, Granato D (2019) Modeling inactivation of Listeria monocytogenes, Shigella sonnei, Byssochlamys fulva and Saccharomyces cerevisiae and ascorbic acid and β-carotene degradation kinetics in tangerine juice by pulsed-thermosonication. LWT 111:612–621

Hodges N (2018) Reproducibility and performance of endospores as biological indicators. In: Microbiological quality assurance. CRC Press, pp 221–233

Ibrahim MT, Briesen H, Först P, Zacharias J (2019) Lethality calculation of particulate liquid foods during aseptic processing. PRO 7:587–607

Li Y, Padilla-Zakour OI (2021) High pressure processing vs. thermal pasteurization of whole Concord grape puree: effect on nutritional value, quality parameters and refrigerated shelf life. Foods 10(11):2608

Manzoor MF, Xu B, Khan S, Shukat R, Ahmad N, Imran M, Korma SA (2021) Impact of high-intensity thermosonication treatment on spinach juice: bioactive compounds, rheological, microbial, and enzymatic activities. Ultrason Sonochem 78:105740

Montanari C, Tabanelli G, Zamagna I, Barbieri F, Gardini A, Ponzetto M, Gardini F (2019) Modeling of yeast thermal resistance and optimization of the pasteurization treatment applied to soft drinks. Int J Food Microbiol 301:1–8. https://doi.org/10.1016/j.ijfoodmicro.2019.04.006

Naczk M, Artyukhova AS (2020) Canning. In: Seafood: resources, nutritional composition, and preservation. CRC Press, pp 181–198

Pereira APM, Stelari HA, Carlin F, Sant'Ana, A. S. (2019) Inactivation kinetics of Bacillus cereus and Geobacillus stearothermophilus spores through roasting of cocoa beans and nibs. LWT 111: 394–400

Rayman Ergün A, Baysal T (2019) Effects of thyme, basil, and garlic oleoresins on the thermal resistance of Bacillus coagulans in tomato sauce. J Food Process Preserv 43(10):e14118. https://doi.org/10.1111/jfpp.14118

Rodríguez-Ramos F, Tabilo EJ, Moraga NO (2021) Modeling inactivation of Clostridium botulinum and vitamin destruction of non-Newtonian liquid-solid food mixtures by convective sterilization in cans. Innovative Food Sci Emerg Technol 73:102762

Setlow P, Johnson EA (2019) Spores and their significance. In: Food microbiology: fundamentals and frontiers. ASM Press, pp 23–63. https://doi.org/10.1128/9781555819972.ch2

Silva FV (2019) Heat assisted HPP for the inactivation of bacteria, moulds and yeasts spores in foods: log reductions and mathematical models. Trends Food Sci Technol 88:143–156. https://doi.org/10.1016/j.tifs.2019.03.016

Soni A, Smith J, Archer R, Gardner A, Tong K, Brightwell G (2020) Development of bacterial spore pouches as a tool to evaluate the sterilization efficiency: a case study with microwave sterilization using Clostridium sporogenes and Geobacillus stearothermophilus. Foods 9(10): 1342. https://doi.org/10.3390/foods9101342

Yanniotis S (2007) Solving Problems in Food Engineering Springer Nature

Zahirinejad S, Hemmati R, Homaei A, Dinari A, Hosseinkhani S, Mohammadi S, Vianello F (2021) Nano-organic supports for enzyme immobilization: scopes and perspectives. Colloids Surf B: Biointerfaces 204:111774. https://doi.org/10.1016/j.colsurfb.2021.111774

Zhang C, Hu C, Sun Y, Zhang X, Wang Y, Fu H et al (2021) Blanching effects of radio frequency heating on enzyme inactivation, physiochemical properties of green peas (Pisum sativum L.) and the underlying mechanism in relation to cellular microstructure. Food Chem 345:128756

Zhao Y, Chen R, Tian E, Liu D, Niu J, Wang W, Zhao Z (2018) Plasma-activated water treatment of fresh beef: bacterial inactivation and effects on quality attributes. IEEE Trans Radiat Plasma Med Sci 4(1):113–120

Zhao Q, Yuan Q, Gao C, Wang X, Zhu B, Wang J, Ma T (2021) Thermosonication combined with natural antimicrobial nisin: a potential technique ensuring microbiological safety and improving the quality parameters of orange juice. Foods 10(8):1851

Zhu Y, Zhang M, Mujumdar AS, Liu Y (2022) Application advantages of new non-thermal technology in juice browning control: a comprehensive review. Food Rev Int:1–22. https://doi.org/10.1080/87559129.2021.2021419

Further Reading

Holdsworth D, Simpson R (2007) Thermal processing of packaged foods. Springer

Chapter 8
Environmental Issues in Food Engineering

Aim The aim of this chapter is to identify key environmental issues associated with food engineering by exploring the effects of food processing and packaging on air, water streams and the land.

8.1 Introduction

The processes and products of food business impact on the natural environment by consuming material and energy resources. It is therefore important to understand how the use of these resources impacts on our environment, so that we can develop strategies for their sustainable use. Ideally, sustainability must be applied across the entire food chain from cradle to grave. However, given that food engineering is primarily applied to product formulation, processing and packaging, we only deal with a segment of the whole chain. In this chapter, therefore, we primarily restrict ourselves to how the ingredients used, processing and packaging impact on the environment, mainly with water and land. Food processing operations do not contribute specifically nor significantly to air pollution, and therefore this aspect has not been addressed. The environmental issues are discussed in terms of engineering water supply and disposal, management of food processing and packaging wastes to reduce their environmental impact by recovering coproducts and energy resources, and the use of life cycle analysis to examine environmental impact.

© Springer Nature Switzerland AG 2022
K. Niranjan, *Engineering Principles for Food Process and Product Realization*,
Food Engineering Series, https://doi.org/10.1007/978-3-031-07570-4_8

8.2   Engineering Water Supply for Food Processing Operations

Water quality and quantity are critical in food processes. Water used in food process either becomes a part of the product, or it can be used as a utility for cleaning, blanching and other purposes. The water is normally drawn into a food factory as potable water delivered from a municipal source. The quality of this water often varies from place to place, but, by and large, the standards for municipal potable water in most countries are drawn-up based on World Health Organisation (WHO) guidelines. The European Union (EU) has established the Drinking Water Directive (98/83/EC) of 1998 pertaining to the quality of potable water.

Given that water quality varies significantly depending on its source, it is not possible to produce consistent quality food products even though the starting materials and the processes employed are the same. Nowhere is this lack of consistency so critical as in the bottled water and beverage industry, where the products contain well over 95 or even over 99% water. It therefore becomes imperative to treat potable water and enhance its quality to a point where the end-product quality becomes independent of the water source. We will first consider the steps involved in the production of potable water from a natural source and then consider how this water is further treated to attain the standards required for use in beverages.

Potable water must have aesthetic, safety and compositional attributes. In terms of aesthetic attributes, the water must be free from colour, taste, and odour and it must also be pleasing – all of which are based on subjective assessments. In terms of safety attributes, it must be free from bacteriological contamination especially free from pathogen; and in terms of compositional attributes, it must be free from natural and anthropogenic (caused by human activity) toxicants e.g. pesticides, fluorides, and heavy metals like arsenic, mercury, cadmium etc. Water from a natural source, such as a river or ground water, is invariably *polluted* and *contaminated* – the latter refers to harmful bacterial presence while the former refers to the presence of organic and inorganic materials – some of which may be suspended in the water and some dissolved in it. Any water treatment process must therefore aim to eliminate *suspended*, *chemical* and *bacterial* loads, and reduce their levels to below legal threshold values. The suspended particles are predominantly colloidal in size although they can also be coarser. The chemical load – as mentioned earlier – may be organic or inorganic; the organic load may be *biodegradable* (i.e. degraded by normal environmental bacteria) or *non-biodegradable* like phenols and conventional plastics. Other forms of chemical load include fertilisers, pesticides and insecticides like endosulfan. Pollution and contamination of natural streams are caused by the mixing of such streams with domestic, industrial and agricultural wastewaters. The purpose of water treatment must therefore be: (1) to remove *turbidity* (largely caused by the suspended matter), colour, taste and odour, (2) to make it safe from pathogens and toxicants.

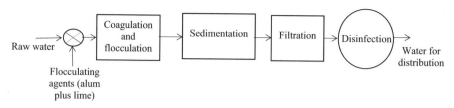

Fig. 8.1 A simplified diagram of a water treatment process

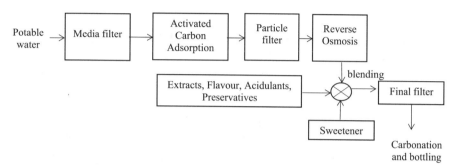

Fig. 8.2 A simplified diagram of water treatment process for soft drink manufacture

Figure 8.1 shows a simple water treatment plant to produce potable water. The first step is to deal with the suspended load, and this is done by adding flocculants which destabilise the colloid surface and promote flocculation. The next step is to let the flocculated suspension into sedimentation tanks for the solids to settle under the influence of gravity. The supernatant from the sedimentation tank also contains suspended solids, but at a significantly reduced load, which is removed by filtration. The filtrate is then disinfected by chlorination, leaving some residual chlorine level to prevent downstream contamination. In some plants, further disinfection is also achieved by ultraviolet radiation.

Water is an active ingredient in many food products. As mentioned earlier, in beverages, its percentage is higher than 95%, while in bottled water, it is the sole ingredient. The quality of potable water obtained from natural streams, as described above, will not meet the product quality standards that is required for brand consistency. For example, chloride, sulphate and other anions negatively affect the flavour of wine, beer, and other beverages, while the flavour composition of other foods can also be affected by the quality of water used. In general, food and beverage manufacturers utilize industrial water purification technologies to further purify potable water into ultra-pure water quality, prior to using the water as an active ingredient. The chlorine level in such waters must typically be kept below 0.1 ppm.

A typical water purification system for soft drink manufacture combines media filtration, reverse osmosis (RO) and ultra-filtration; see Fig. 8.2. Media filters aim to reduce the *silt density index* by lowering suspended solid loading for particles in the range 10–20 μm and above. Activated carbon filters remove off-tastes, odours, chlorine, chloramines, low molecular-weight organics and trihalomethanes (THM). Reverse Osmosis (RO) effectively rejects monovalent and divalent ions (e.g. alkalinity, sodium, chloride, calcium, magnesium, etc.), viruses, bacteria, and pyrogens.

8.3   Characterisation of Wastewater and Its Impact on Natural Water Streams

Food processing facilities use water for conveying product, washing, processing and cleaning operations. The water which has gone through a process and has got mixed with other water streams is known as *wastewater*. The volumes and pollutant loads are invariably very high. The nature of pollutants and contaminants in wastewater is similar to that present in natural streams, i.e. wastewater also contains *suspended*, *chemical* and *bacterial* loads as mentioned in Sect. 8.1, but their levels are so significantly higher, that such streams, if directly discharged into natural water streams, could threaten the very existence of aquatic life and disturb the ability of nature to regenerate itself. It is therefore imperative that wastewater is treated and rendered safe before discharging into natural streams.

Before considering treatment methods, it is important to identify and characterise the pollutants and contaminants present in wastewater. Like in the case of raw water, wastewater contains suspended solids which vary in dimension from coarse particles in the micron range through to colloidal particles in the nano meter range. The chemical contaminants can be divided into organic and inorganic matter. The organic matter can further be classified into biodegradable and non-biodegradable matter. Chemicals containing nitrogen and phosphorus are also present, which can potentially act as nutrients for the bacteria and consume dissolved oxygen in the process. In addition, wastewater can also contain grease and fat, volatile compounds and dissolved gases which often cause offensive odours.

The key measures of the contaminants in water are: (1) Oxygen demand (biological and chemical), (2) Indicator microbial loads, (3) Solids content (suspended and dissolved), (4) Grease and fat particles and drops, (5) Chemical analyses (ammonia and nitrate concentrations, total and reactive phosphorus, pH and alkalinity) and (6) Volatile compounds including dissolved gases and odour causing compounds like H_2S.

The Biochemical Oxygen Demand or BOD is the mass of oxygen utilized in milligram by a mixed population of microorganisms during aerobic oxidation of the organic matter present in 1 l of wastewater at a controlled temperature of 20 °C. It is also interesting to note that mg/L is also the same as parts per million or ppm, and the two are interchangeable. In theory, it would take an infinitely long time for the microorganisms to degrade all the organic matter present in a given volume of wastewater. Thus, the measured *cumulative value* of BOD will keep on increasing with time, until it reaches, what is commonly known as the *ultimate* BOD (BOD_∞), when all the nutrients in the wastewater are virtually exhausted. If BOD_t is the cumulative oxygen demand up to a time t expressed in days, the rate at which BOD_t increases with time, to a reasonable approximation, may be assumed to be proportional to $(BOD_\infty - BOD_t)$. In other words, the cumulative value of BOD increases with time at a rate that depends on how close its value is to the ultimate BOD, i.e. the rate is high initially, but slows down as the BOD value approaches BOD_∞. Mathematically, the differential equation representing the rate of change of cumulative BOD_t can be written as:

$$\frac{d\,BOD_t}{dt} = k(BOD_\infty - BOD_t) \tag{8.1}$$

where k is the constant of proportionality. At the start of the biological degradation, $BOD_t = 0$, so BOD_t at any time is given by solving Eq. 8.1 with this initial condition to yield:

$$BOD_t = BOD_\infty\left(1 - e^{-kt}\right) \tag{8.2}$$

The BOD value is therefore time dependent. It is customary to determine BOD over a 5-day period (i.e. BOD_5) when 60–70% of the organic matter, mainly organic substrates containing carbon, hydrogen and oxygen, is oxidized. Food processing wastewater also contains proteins and other nitrogenous matter, so the nitrifying bacteria will also exert a measurable oxygen demand, but this normally occurs after 6–7 days. The organic nitrogen is microbiologically converted into NH_3/NH_4^+ which in turn undergo biological nitrification to form NO_2^{-1} (nitrite), and eventually, NO_3^{-1} (nitrate). Equation 8.2 can be used to determine, say, BOD_8, if BOD_5 and the rate constant are known. Typical value of k varies between 0.1 and 0.6 day^{-1} at 20 °C. The value of k is clearly temperature dependent and its variation with temperature is given by an Arrhenius relationship (see Chap. 5, Eq. 5.8). In the case of wastewater treatment, the variation of k with temperature is approximately given by:

$$k_{T_2} = k_{T_1}\theta^{T_2 - T_1} \tag{8.3}$$

where $\theta = 1.135$ for $4 < T < 20$ °C, and $\theta = 1.056$ for $T > 20$ °C. Alternatively, if k_{20} i.e. the value of k at 20 °C, is known – which is normally the case, then $k_T = k_{20}1.047^{T-20}$.

In practice, specialized BOD bottles with air-tight seals, which are designed to allow full filling with no air space, are used. The bottles are filled with the sample to be tested and the concentration of dissolved oxygen in mg/L is measured using a dissolved oxygen probe (DO probe) at the start, and after, say, 5 days of storage in a dark incubator at 20 °C. The BOD is calculated as the difference in the dissolved oxygen concentrations, which represents the oxygen consumed by the respiring microorganisms feeding on the organic nutrients. The ultimate BOD is normally determined in terms of the DO depleted over 60 days. If the water is highly polluted, it may be necessary to dilute the samples with de-ionised water (DI water), and the BOD can be determined after correcting for dilution.

The Chemical Oxygen Demand (COD) represents the oxygen required (again in mg per litre of the wastewater sample) to chemically oxidise all the organics present in the sample. It is the total amount of oxidisable organics (biodegradable and nonbiodegradable and both dissolved and particulate), measured by the amount of oxygen in the form of oxidising agent required for the oxidation of organic matters, by heating the sample in concentrated sulphuric acid (50%) containing potassium dichromate. The amount of dichromate consumed is reflected in a colour change from orange through to green, which is determined using a colorimeter. It is therefore possible to determine COD much more rapidly than BOD_5. The value of COD will obviously be greater than the value of BOD_5, but for a given wastewater, the ratio of the two is generally constant. In the case of food processing wastewater, the COD is generally twice BOD_5. Thus, when an average COD:BOD_5 ratio has been established for a given sample of wastewater, then a relatively simple and quick COD test can be used to predict BOD with relative reliability.

Nowadays the total organic carbon (TOC) testing is gaining popularity because of the availability of the instruments which can measure this parameter with relative ease. The measurement of TOC is based on the oxidation of organics present in water to carbon dioxide, which can be achieved either (1) photo-catalytically (i.e. conversion of organics into CO_2 by ultra-violet or UV radiation), or (2) chemically (using persulfate in a UV-irradiated chamber) or (3) by high-temperature combustion. It is also possible to establish empirical relationships between TOC and BOD_5 for a given wastewater.

The term oil and grease (O&G) has replaced the older term FOG (fat, oil and grease) and can come into wastewater streams from the plant and animal origin materials used in food processing. These are hydrophobic materials and have a low solubility in wastewater, therefore are prone to possess relatively low biodegradability by microorganisms. These materials become more soluble in wastewater at higher temperatures and will form emulsions at lower temperatures. Oil and grease are notorious for causing blockages in pipelines and pump failures. The amount of oil and grease in a sample of wastewater is usually determined by acidifying the water to a pH of 2 using hydrochloric or sulfuric acid, and then extracting the oil and grease in an organic solvent, normally hexane. The mass of oil and grease extracted is determined gravimetrically after distilling off the solvent.

The Total Suspended Solids (TSS) is the mass of dry suspended particles in a given mass of wastewater, that can be trapped by a filter using a filtration apparatus. The level of suspended solids in wastewater can influence the method used to treat the water. It can be accurately measured by filtering a known volume of a sample of wastewater, drying the filter and captured solids, then weighing the filter to determine the weight of the captured suspended solids in the sample. Such measurements can only be undertaken off-line. There are instruments available that can measure this parameter on-line in terms of turbidity units – measured as the extent of light scatter – which can be calibrated against the levels of suspended solids.

The term *nutrient pollution* – an apparent contradiction in terms – is used to highlight excessive level of nitrogen and phosphorus present in water, which is harmful to human health as well as to the environment. The single largest source of nutrient pollution is agricultural fertilizers, while storm water run-off also contributes significantly to this pollution. Nitrogen and phosphorus are present in wastewater in different forms. Ammonia in equilibrium with ammonium ion (NH_3/NH_4^+), organic nitrogen, nitrite and nitrate are four convertible forms of nitrogen. Phosphorus can be present as phosphate, orthophosphate, organic phosphate, and condensed phosphate (metaphosphate, polyphosphate, pyrophosphate). Increasingly, regulatory authorities all over the world are specifying threshold levels of the various forms of N and P that can be discharged after treatment, which has made it imperative to measure these values, normally be resorting to instrumental methods based on techniques such as ion chromatography combined with colorimetric methods.

Urea, food processing wastes and chemical cleaning agents all contribute to Nitrogen in wastewater. As mentioned above, the nitrogen is present in water in different forms: (1) Ammonia Nitrogen (NH_3-N or NH_4^+-N), (2) Nitrite Nitrogen (NO_2^- - N) and (3) Nitrate Nitrogen (NO_3^-- N). Nitrites and nitrates are formed in the wastewater during the biological process. In addition, some nitrogen will be present in a dissolved state (e.g. proteins from food processing wastes) as well as associated with the biomass present in the wastewater; this nitrogen is known as organic-N. It may be noted that all organic –N cannot be removed by the biological processes occurring during wastewater treatment. The sum of all types of nitrogen is the total nitrogen. The organic nitrogen is normally determined by a method developed by Johan Kjeldahl, and is also known as the Kjeldahl N. The difference between total nitrogen and the Kjeldahl N gives the total inorganic nitrogen (ammonia/ammonium, nitrites and nitrates). In water, ammonia is normally in the form of ammonium ion (especially when the pH is between 6 and 9). Autotrophic bacteria (i.e. those bacteria which can obtain their carbon needs from non-organic sources), biologically convert ammonium into nitrites and then nitrates; and consume oxygen while doing so. In other words, nitrifiers also contribute to the BOD. Autotrophs grow much more slowly than heterotrophs and take time to maintain their balance in the wastewater, but higher temperatures, higher DO concentrations and a pH between 6.8 and 7.5 promote their growth.

$$2NH_3 + 3O_2 \rightarrow 2NO_2^- + 2H^+ + 2H_2O$$
$$2NO_2^- + O_2 \rightarrow 2NO_3^-$$

Adding the two above equations, we have: $2NH_3 + 4O_2 \rightarrow 2NO_3^- + 2H^+ + 2H_2O$

In other words, 1 kg of ammonia-N removal requires $64/14 = 4.6$ kg oxygen to convert to nitrate. Some of the heterotrophs present in water can absorb oxygen from the nitrites and nitrates under anoxic conditions for respiration, and in the process, produce nitrogen gas which bubbles out of the water

Table 8.1 Typical values of BOD$_5$ and suspended solid levels (SS) in wastewaters from different sectors of food processing: These values are only indicative and can vary significantly depending on the processing methods and protocols applied at any given factory site

Wastewater source	BOD$_5$ (mg L^{-1})	SS (mg L^{-1})
Carbonated beverages	400–600	50–70
Brewery	1500–1700	65
Rice wash	700–900	150
Vegetable oil	8000	1700
Dairy	300–800	200–400
Sugar beet	450–2000	800–1000
Slaughterhouse	1500–2500	800

Thus, wastewater is normally characterised in terms of Suspended solids, BOD, COD, TOC, N and P content, level of dissolved oxygen, pH, total chlorine and microbial coliform loading. It is necessary to note that these characteristics depend on the source of the wastewater and there is no uniformity across the various food processing sectors, as well as within any given sector. In addition, the volumes of wastewater to be treated also vary depending on factory throughput and protocols employed. Table 8.1 gives indicative values of BOD and suspended solid contents.

Problem 8.1 In order to determine BOD of a wastewater, 10 ml of the wastewater is diluted to 300 ml using distilled water and the dissolved oxygen concentration was noted initially to be 8.5 mg/l, which became 5 mg/l after 5 days. Determine BOD$_5$ of wastewater and also estimate the ultimate BOD if the rate constant $k = 0.23d^{-1}$.

The amount of oxygen consumed over a 5 day period by the diluted waste is $(8.5 - 5) = 3.5$ mgL^{-1}. Since the wastewater is diluted 30 times, the BOD$_5$ of the wastewater is: $3.5 \times 30 = 105$ mgL^{-1}. From Eq. (8.1), BOD$_5$ = BOD$_\infty(1 - e^{-0.23 \times 5})$, whence, BOD$_\infty = 153.65$ mgL^{-1}.

Problem 8.2 Determine carbonaceous BOD$_{10}$ from the following experimental data obtained on a sample of wastewater:

Analysis	Day 0 (mg L^{-1})	Day 10 (mg L^{-1})
DO	8.3	1.4
NH$_3$-N	2.1	0.8
NO$_2$-N	0	0.1
NO$_3$-N	0.5	2.1

Total BOD$_{10}$ = $8.3 - 1.4 = 6.9$ mgL^{-1}. The nitrogenous oxygen demand must now be calculated from the relevant data in the Table. We know that Ammonia undergoes nitrification to form nitrite and then nitrate. The amount of nitrite-Nitrogen = 0.1 mgL^{-1}, equivalent to 0.33 mgL^{-1} of nitrite mass. The stoichiometric formation of nitrite, as mentioned earlier, can be represented as:

$$2NH_3 + 3O_2 \rightarrow 2NO_2^- + 2H^- + 2H_2O$$

Thus, 92 mg L^{-1} Nitrite has a BOD of 96 mgL^{-1} Hence, 0.33 mgL^{-1} nitrite has a BOD of 0.34 mgL^{-1}.

Likewise, 1.6 mgL^{-1} NO_3-N is equivalent to $(62/14) \times 1.6 = 7.09$ mgL^{-1} of nitrate mass. The stoichiometric formation of Nitrate, as already mentioned, is represented as:

$$2NO_2^- + O_2 \rightarrow 2NO_3^-$$

Thus 124 mg L^{-1} nitrate is equivalent to BOD of 32 mgL^{-1}. Therefore, 7.09 mgL^{-1} is equivalent to 1.83 mgL^{-1}.

Thus, the total nitrogenous BOD $= 0.34 + 1.83 = 2.17$ mgL^{-1}.

The carbonaceous BOD is therefore 6.9–$2.17 = 4.73$ mgL^{-1}.

Problem 8.3 Wastewater from a processing plant having a carbonaceous BOD of 300 mgL^{-1} enters, at a flow rate of 8×10^{-3} m³s^{-1}, a continuous stirred tank reactor (CSTR) containing a suspension culture activated sludge biomass, where it is treated. If the mean residence time in the reactor is a day, and the BOD reduction can be considered to be a first order process with a rate constant of 1 d^{-1}, what is the depth of liquid in the CSTR if its diameter is 15 m, and what is the BOD of the treated wastewater leaving the reactor?

Solution If Q is the volumetric flow rate of the wastewater and V is the reactor volume, the mean residence time $\tau = V/Q$. With $\tau = 1$ day $= 86{,}400$ s and $Q = 8 \times 10^{-3}$ m³s^{-1}, $V = 691.2$ m³. With diameter being 15 m, the height is: $V/[(\pi/4)d^2 = 3.91$ m

We can assume that a mass balance of the carbonaceous pollutants, expressed in terms of BOD, to be valid around the CSTR

$$Q(BOD)_i = Q(BOD)_o + rV \tag{8.4}$$

where $(BOD)_i$ is the inflow BOD, $(BOD)_o$ is the BOD inside the CSTR which is also the exit BOD, and r is the rate of drop of BOD per unit volume $= k(BOD)_o$ where $k = 1d^{-1}$. It may be noted that dissolved oxygen is also a reactant. However, it's influence on the rate of reaction is assumed to be zero order (i.e. the reaction rate is independent of the DO level). Dividing Eq. (8.4) by Q and noting that $\tau = V/Q = 1$ day, we have:

$$(BOD)_o = (BOD)_i/[1 + k\tau] = 200 \times 10^{-3}/[1 + 1]$$
$$= 100 \times 10^{-3} \text{ kg m}^{-3} \text{ or } 100 \text{ mg}L^{-1}.$$

8.4   Wastewater Treatment and Disposal

The broad objective of wastewater treatment is to lower to acceptable discharge standards the pollutants and contaminants present in wastewater which include suspended solids and fat, and organic matter including nutrients containing nitrogen and phosphorus. In addition, the pH may also have to be adjusted in some cases. The treatment and subsequent discharge standards for wastewaters are set by governments in all countries. For example, within the European Union, "The Urban Waste Water Treatment Directive (91/271/EEC)" of 1991, and its subsequent updates pertain to discharges of municipal and some industrial wastewaters. As a rule of thumb, the BOD value must be reduced from the values stated in Table 8.1 to less than 20 mg L^{-1} in the treated effluent while the suspended solids must also be less than 20 mg L^{-1}.

Suspended particles are normally removed by gravity and/or filtration, while most other organics are removed by growing microorganisms which use the organics as nutrients. With stringent regulatory compliance requirements, the cost of treating wastewater is increasing, and every effort must be made to lower the costs. Preventative approaches are far better than curative approaches. It is therefore important that the design methodologies employed for engineering processes and products include water reuse and recycling, so that the net requirement of fresh water, and the amount of wastewater discharged into natural streams, are both minimised. The methods adopted to treat wastewater are discussed below.

As far as possible, every effort must be made to stop oil and grease at source. Grease interceptors or grease traps are available for this task, which collect the grease. It is however important to empty and clean the grease interceptors regularly. At the wastewater treatment plant, oil and grease can be removed by bioaugmentation which requires the regular injection of specialised organisms to feed on the fat, especially since the introduced organisms do not tend to keep pace with a system's main bacterial population. Newer methods, however, focus on stimulating the main bacterial population to break down oil and grease, by utilising optimised fermentation-based yeast proteins, micro-nutrients and specialised surfactants to stimulate the indigenous bacterial population.

In general, wastewater treatment facilities can be divided into three sections: (1) Primary treatment, mainly designed to remove suspended solids and fat drops by a combination of settling, screening and dissolved air floatation (2) Secondary treatment, at the heart of which lies the consumption of the carbonaceous substrates by natural environmental microorganisms, and (3) Tertiary treatment which refers to the removal of specific constituents from the wastewater which cannot be removed by secondary treatment. Secondary treatment involving the biological degradation of carbonaceous substrates can be achieved *aerobically by, say, using activated sludge* (AS) or by *anaerobic digestion* (AD). In aerobic processes, the carbon substrate is converted to carbon dioxide and water by the microorganisms in the presence of dissolved oxygen, which also ends up producing more microorganisms in the process. The process therefore relies on the continuous addition of oxygen into the wastewater. Typically, half the BOD is converted to CO_2 and H_2O, while the other half is converted into biomass. The biomass produced thus is very significant in mass and it must be utilised. Possible options are its use as a soil conditioner or as a fuel after dehydration. Another possible option may be to further digest it anaerobically. It is estimated that the cost of biomass disposal, which can be very significant especially if land is needed to dispose it, can amount to almost half the cost of wastewater treatment.

Anaerobic digestion involves the microbial consumption of the carbonaceous substrates present in either the above biomass, or the wastewater itself, in the virtual absence of oxygen to form methane and carbon dioxide – which is known as biogas and has fuel value. The conversion of the substrates occurs in two main stages with the dissolved carbon substrate being first converted to organic acids by facultative microbes followed by the conversion of the acids into methane by anaerobes. Anaerobic digestion is cheaper than aerobic process to operate, mainly because there is no need to continuously sparge air or oxygen. From an environmental point of view, this process helps in reduction of solids in sludge which leads to a reduction in the final volumes of treated waste to be disposed. The disadvantages of AD are that it requires a high level of investment in large tanks and other process vessels, and it does not convert as large a proportion of the carbon in the biomass into biogas.

Figure 8.3 shows a simplified diagram of the processes used to treat wastewater emanating from a processing factory. Normally, the wastewater collected in a sump, is pumped first through a screen in order to remove floating matter as well as large objects. The wastewater then enters a grit chamber, primarily to be rid of inorganic solids which have a higher material density of 2650 kg m^{-3} than organic solids. Of

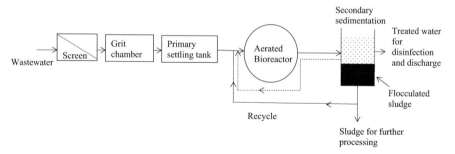

Fig. 8.3 A simplified schematic diagram of the key steps involved in treating wastewater. Biological treatment of the water occurs in the aerated bioreactor where the biomass can be a *mixed suspension culture* or be *immobilised* on solid matrices. In activated sludge process – the most commonly employed process in practice – the biomass is suspended in the reactor, and a part of the flocculated sludge collected in the secondary sedimentation (around 30%) is recycled in order to maintain an adequate population of active microorganisms to cause biodegradation of the organic wastes. In *immobilised* bioreactors, e.g. trickle bed reactor or rotating disc contactor, the biomass is fixed on the solid support inside the bioreactor and there is no need to recycle the flocculated biomass. Instead, some of the supernatant may be recycled (shown by the dashed line) mainly to distribute and irrigate the wastewater uniformly over the immobilised microorganisms

course, the bulk density of inorganic materials like sand and mud, in general, will be lower at around 1650 kg m^{-3}. The water then enters the primary sedimentation tank where inorganic particles and some organic matter settle down. About 60% of suspended solids are removed and an equivalent of 30% BOD is lowered. The wastewater then enters the Biological Wastewater Treatment chamber, shown as *aerated bioreactor* in Fig. 8.3. The wastewater still contains suspended, colloidal and dissolved organic solids, and possibly some inorganic solids, but at a significantly lower concentration than at the start. Biological treatment aims to remove organic solids which become nutrients for aerobic microbial activity, and results in CO_2, water, and more flocculating biomass which subsequently settle down in the secondary sedimentation unit. Thus, most of the biodegradable organic matter in the wastewater has been converted into flocculating biomass, which can be recovered as the sludge. The supernatant liquid, free of most suspended solids and most of the organic dissolved solids, is then disinfected by chlorination or UV treatment to remove pathogens.

In practice, the primary treatment units (i.e. the units placed before the aerated bioreactor) may also include a comminutor or grinder as well as equalisation tank and flow measuring instrument. The equalisation tank collects the wastewater, so that it can be pumped downstream at a uniform rate which is not affected by flow fluctuations caused by over- or under-supply of wastewater at different times.

The aerated bioreactor shown in Fig. 8.3 can either be a mixed suspension bioreactor where the flocculating cells are formed which are freely "swimming" or dispersed in the wastewater as it is being treated. Clearly, the suspension in the bioreactor is also exiting the reactor under steady state conditions, and it may be necessary to recycle a part of the flocculated cells after sedimentation in order to maintain a healthy population of biomass in the reactor. An alternative method, which avoids significant loss or a washout of biomass from the reactor, is to use a *trickle bed filter* where the cells are immobilised on a solid support made from ceramic or plastic and the incoming wastewater is uniformly distributed over the solids. The trickling liquid gets aerated as it flows over the solids and the additional biomass formed also adheres to the solid support to form a growing biofilm. A second method of immobilising the cells is to do so on plastic discs (polythene or PVC), which are mounted on a rotating shaft turning at 2–4 revolutions per minute, in such a way that the discs are partially submerged in the wastewater. As the discs rotate, they pick up a film of wastewater when submerged, which gets exposed to air during the unsubmerged part of the rotation and allows aerobic microbial degradation to occur. Approximately 95% of the surface area is thus alternately submerged in wastewater and then exposed to the atmosphere above the liquid. This bioreactor configuration is known as the *rotating biological contactor* (RBC).

It is necessary to note that Fig. 8.3 essentially traces the flow of wastewater through the treatment system, but it does not highlight what happens to the other streams which are formed. For instance, thick solid suspensions will emanate from the screen, grit chamber and the primary settling tank. It would be instructive to reflect on the nature and composition of these suspensions and how these can be utilised as a resource and/or be safely disposed.

8.5 Food Sustainability

This term is very confusing because it means different things to different groups of individuals. For the purpose of understanding food sustainability from an engineering or processing perspective, we refer to the United States Environmental Protection Agency (EPA) which addresses *Sustainable Management of Food as a systematic approach that seeks to reduce wasted food and its associated impacts over the entire life cycle, starting with the use of natural resources, manufacturing, sales, and consumption and ending with decisions on recovery or final disposal.* The EPA makes a subtle difference between "wasted food" and "food waste", with the former meaning food that was not used for its intended purpose (and therefore wasted), while the latter refers to food that no longer has value and needs to be managed as waste. Indeed food waste refers to plate waste (i.e food that has been served but not eaten), spoiled food, or peels and rinds, bones etc considered inedible and sent to feed animals, to be composted or anaerobically digested, or to be landfilled or combusted with energy recovery. The EPA also recommends a food hierarchy (Fig. 8.4) where the upper layers provide the highest benefits to environment, society and the economy.

At this stage, it would be instructive to consider the characteristics of food wastes. The quality and quantity of wastes produced by different sectors of the food industry

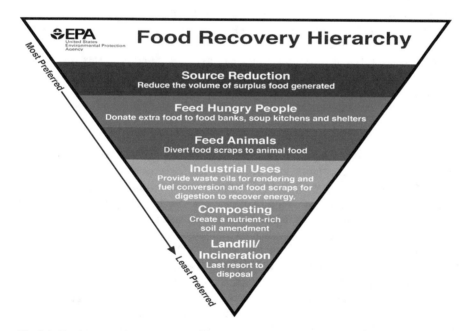

Fig. 8.4 Food recovery hierarchy developed to prevent and divert wasted food. The same hierarchy can be used to prevent and divert food wastes except with the deletion of the second layer from the top i.e. "Feed Hungry people". (Reproduced with permission: https://www.epa.gov/sites/production/files/2019-11/food_recovery_hierarchy_-_eng_high_res_v2.jpg)

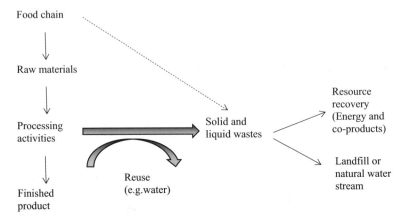

Fig. 8.5 The relationship between food hierarchy and food processing wastes (Waldron 2007)

vary depending on the type of food being processed. As a first approximation, food waste levels can be estimated from mass balances as the difference between the mass of raw materials and water entering a processing facility, and the mass of the finished products leaving it. In general, even though the percentage loss, estimated thus, may be low, the quantities of wastes produced, and hence their environmental impacts, are very high. Food processing operations produce many varied types of wastes that can be categorized into solid, liquid and gaseous wastes. But this is an oversimplification because the solid wastes invariably contain water, the so-called liquid wastes contain suspended and dissolved solids, and even the gaseous wastes can contain dust and mist. Thus, food wastes must be treated as multi-phase and multi-component systems. It is always possible to recover resources – either in the form of co-products or as energy from solid and liquid wastes. The relationship between food processing wastes and the food hierarchy given in Fig. 8.4 can be illustrated as shown in Fig. 8.5.

Resource recovery can also include chemical transformation of wastes and one of the most mentioned illustration in this regard is the conversion of used frying oil into diesel. In principle, frying oil can be used in internal combustion engines, but its viscosity is very high; it contains free fatty acids; and undergoes oxidation and polymerisation reactions to form gums which can leave carbon deposits in engines. However, the triglycerides in frying oil can be transesterified using an organic alcohol (usually methanol, which has a low cost) to produce fatty acid esters (commonly known as biodiesel) and glycerol. The characteristics of biodiesel are like conventional diesel especially in terms of viscosity and cetane rating (which is a measure of the quality or performance as a fuel in diesel engines). The transesterification reaction requires catalysis either by an acid or alkali. Alternatively, enzymes (lipase) can also be used to catalyse this biotransformation. After transesterification, the reaction mixture separates into two phases: a lighter phase containing the esterified product and a heavier phase containing glycerol. Both phases will contain unreacted methanol which has to be recovered for reuse. The esterified phase is purified for use as fuel, while the glycerol phase also undergoes purification for recovering glycerol as a co-product. This process also generates wastewater which will require treatment.

8.6 Food Packaging and Sustainability

Packaging is an integral part of food processing. Without packaging, it would be impossible to assure product quality over any stipulated period, and the quantity of wasted foods will mount to unjustifiable levels. In general, a good food packaging must serve to contain, preserve, protect, dispense, promote and inform. In other words, it must possess very specific functionalities. But it is also necessary to recognise that the contents are consumed; not the packaging. So packaging poses considerable disposal problems at the end of use which has an environmental effect. This effect is further exasperated by the fact that there is a grotesque disparity between the "life spans" of the food and packaging. While the food can last from anywhere between a few days up to a few months, the packaging materials used currently are sourced unsustainably and seem to last to eternity! It is therefore necessary to make food packaging sustainable whilst endeavouring to reduce the disparity between the life spans of the food and its packaging.

The market share of currently used packaging materials is given in Table 8.2. It is clear from the table that plastics – made from petrochemical sources – contribute almost 37%. Not only is this source unsustainable, but the damage to the environment caused by plastic packaging, after use, is incalculable and not yet been fully ascertained. The key question therefore is whether this material ought to be replaced? Or managed better after use? While opinions are divided, there is a strong focus on developing food packaging from more sustainable materials which aim to replace plastics. These materials are either *biodegradable* or *compostable.*

The food hierarchy helps to identify the options available in respect of waste handling. However, it does not provide us with a methodology to establish which of the options is more environmentally benign. A quantitative tool which can be used to compare the environmental impacts, for instance, of using different raw materials and processes to manufacture a given product is the Life Cycle Analysis which is discussed below.

Table 8.2 Market share of packaging material

Packaging material	Market share (%)
Paper and board	34
Rigid plastic	27
Flexible plastic	10
Glass	11
Beverage cans (including aluminium)	6
Other metals	9
Other materials	3

https://www.foodpackagingforum.org/food-packaging-health/food-packaging-materials

8.7 Life Cycle Analysis

As indicated earlier, human activities, inevitably, have an impact on the environment, mainly through effects on land, natural water streams and air. Life Cycle Analysis or Assessment (LCA) of a manufactured product is a technique used to assess *the environmental impacts of each and every process involved in the life of a product* starting from its raw materials extraction and processing, through to their transformation into the product, followed by its use, and finally, its fate after use. Thus, it represents a *cradle to grave* analysis of the environmental impact of the product. In the case of a plant-based food product, for instance, the LCA will include the environmental impact of:

1. Growing the raw materials and other ingredients used in product formulation,
2. Transportation of these materials to the processing site,
3. Water and energy use in processing,
4. Packaging material and sourcing characteristics,
5. Management of wastes generated in the processing site
6. Transportation and maintenance of the product, and its distribution to consumers, and
7. The fate of product and packaging post use.

According to ISO 14040 and 14044, the four highly interactive stages of an LCA, illustrated in Fig. 8.6, are: (1) Goal and scope definition, (2) Inventory analysis, (3) Impact assessment and (4) Interpretation. The definition of *goal and scope* is absolutely critical because the analysis is based on the statements made in this stage, which states the purpose of the study, defines the product, the system boundaries and assumptions. The *inventory analysis* is at the heart of LCA and it requires the inputs to the process and products, the actual processing methods employed as well as the detailed outputs. A typical inventory analysis is schematically shown in Fig. 8.7. The *impact assessment* aims to classify and characterise the environmental impacts into

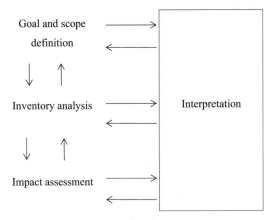

Fig. 8.6 Four stages of life cycle assessment

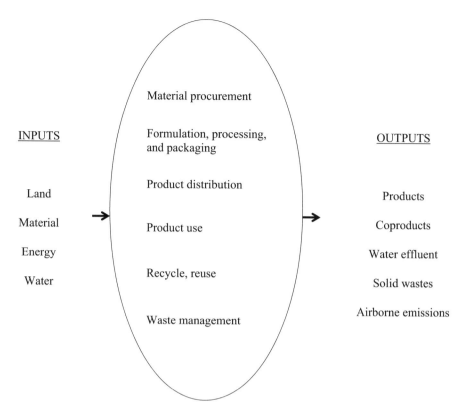

Fig. 8.7 Life cycle inventory analysis

groups and normalise these for comparison. The impact categories include global impact (e.g. global warming and ozone depletion), regional impact (acidification and eutrophication), and local effects (odour nuisance, and hazardous wastes etc.). The normalisation for global warming potential (GWP) of a greenhouse gas, for instance, is done with respect to carbon dioxide which is taken as 1. The final interpretation stage of LCA is where the inventory and impact assessment results are discussed together, and significant environmental issues are identified to draw conclusions and recommendations which are consistent with the goal and scope of the study.

The advantages of LCA are: (1) it allows the analysis of all the steps within the life cycle of a food product; (2) it enables quantitative comparisons to be drawn between different alternative raw material sourcing, manufacturing and packaging options, so that the most sustainable approach may be adopted; (3) it helps in corporate and regulatory policy formulation; and (4) it can also be used as a marketing tool. The key disadvantages include: (1) All data for inventory analysis are often not be readily available and several assumptions have to be made which makes the analysis questionable especially when the LCA scope is defined very broadly, (2) the data and conclusions can also be very time sensitive, and (3) the

Table 8.3 Some examples of Life Cycle Analysis performed on food systems

Food system	Major environmental impact	References
Bread	Primary wheat production and transportation	Kulak et al. (2015)
Beer	Wort production, filtration, packaging	Amienyo and Azapagic (2016)
Tomato ketchup	Processing and packaging	Anderson et al. (1998)
Milk	Primary farming	Thomassen et al. (2008)
Beef	Pastures, energy and manure storage	Huerta et al. (2016)
Rice	Depended strongly on farm size and practices	Habibi et al. (2019)

analyses are not globally harmonised and there is no single global standard for LCA. Yet there are a number of LCAs that have been undertaken for food products and Table 8.3 lists key references.

References

Amienyo A, Azapagic A (2016) Life cycle environmental impacts and costs of beer production and consumption in the UK. Int J Life Cycle Assess 21:492–509. https://doi.org/10.1007/s11367-016-1028-6

Andersson K, Ohlsson T, Olsson P (1998) Screening life cycle assessment (LCA) of tomato ketchup: a case study. J Clean Prod 6:277–288

Habibi E, Niknejad Y, Fallah H, Dastan S, Tari DB (2019) Life cycle assessment of rice production systems in different paddy field size levels in north of Iran. Environ Monit Assess 191:202. https://doi.org/10.1007/s10661-019-7344-0

Huerta AR, Guereca LP, Rubio Lozano MS (2016) Environmental impact of beef production in Mexico through life cycle assessment. Resour Conserv Recycl 109:44–53. https://doi.org/10.1016/j.resconrec.2016.01.020

Kulak M, Nemecek T, Frossard E, Chable V, Gaillard G (2015) Life cycle assessment of bread from several alternative food networks in Europe. J Clean Prod 90:104–113

Thomassen MA, van Calker KJ, Smits MCJ, Iepema GL, de Boer IJM (2008) Life cycle assessment of conventional and organic milk production in the Netherlands. Agr Syst 96:95–107. https://doi.org/10.1016/j.agsy.2007.06.001

Waldron KW (2007) Handbook of waste Management and Co-product Recovery in Food Processing, Volume 1, p.16 (Edited by K.W. Waldron) Woodhead Publishing (Cambridge, England)

Further Reading

Anastopoulos G, Zannikou Y, Stournas S, Kalligeros S (2009) Transesterification of vegetable oils with ethanol and characterization of the key fuel properties of ethyl esters. Energies 2:362–376. https://doi.org/10.3390/en20200362

Roy P, Nei D, Orikasa T, Xu Q, Okadome H, Nakamura N, Shiina T (2009) A review of life cycle assessment (LCA) on some food products. J Food Eng 90:1–10. https://doi.org/10.1016/j.jfoodeng.2008.06.016

Chapter 9
An Engineering View of the Fate of Food in the Gastrointestinal Tract (GIT)

Aim The principal aims of this chapter are to understand the key parts of the gastrointestinal tract from the oral cavity to the colon, and gain insights into the flow, mixing, structural and biochemical changes occurring in the food and drink consumed. This chapter also throws light on some of the mathematical modelling work done in relation to the flow of foods through the stomach, and the flow and biochemical transformations occurring in the small intestine and the colon.

9.1 Introduction

All the earlier chapters in this book dealt with engineering principles relevant to producing safe food products possessing adequate keeping and eating qualities. In other words, these chapters dealt with engineering principles that applied *from farm to the fork*. With increasing focus on the effects of food on health, especially the links between diet and disease and the roles that food can potentially play in preventative as well as curative aspects of ill health, it has become imperative for engineers to gain insights into the fate of food post consumption, and use these insights to design processes and products which have a favourable impact on human health. This chapter therefore focuses on the structural and chemical changes occurring in different parts of the gastrointestinal tract, particularly the mouth, stomach, small intestine and the colon. The emphasis is not on anatomy and physiology, but instead, on the flow, mixing, mass transfer and biochemical reaction aspects which bring about structural and chemical changes in the food as it moves through the tract. Such analyses will help to design processes for producing products which will not merely be utilised for body functions, but be utilised at the right pace so as to strike the right balance between palatability, satiety, digestibility (i.e. processing of food in the GI tract to absorb useful components, produce energy and eliminate wastes) and metabolism (how the cells utilize the energy absorbed from food during digestion). In addition, it is also necessary to recognise that not all the food consumed by an

© Springer Nature Switzerland AG 2022
K. Niranjan, *Engineering Principles for Food Process and Product Realization*,
Food Engineering Series, https://doi.org/10.1007/978-3-031-07570-4_9

individual is assimilated and utilised for storage and metabolic activities. *Bioavailability* expresses the fraction of ingested nutrient that reaches the systemic circulation and is ultimately utilized. Before becoming bioavailable, bioactive compounds must also be released from the food matrix and be modified in the GI tract. *Bioaccessibility* is defined as the quantity of a compound that is released from its matrix in the gastrointestinal tract, which becomes available for absorption (e.g. enters the blood stream). Thus, bioaccessibility is the first step to bioavailability (Rein et al. 2013). A detailed understanding of the structure, function and the mechanisms by which different parts of the GI tract act will enable us to gain insights into the bioavailability and bioaccessibility of the nutrients in food.

9.2   The Gastrointestinal Tract (GIT)

The digestive tract in humans has evolved to efficiently extract the optimum amount of nutrients and other bioactives from the food consumed, whilst at the same time reject pathogens and toxic substances. The GIT is therefore made up of compartments or chambers which have very specific functions, that run in a highly coordinated and controlled manner with the help of the nervous system. The main parts of the GIT are: (1) the *mouth*, which interacts with the food's sensory properties and undertakes what is commonly known as *oral processing*; (2) the *stomach* which is also called the gastric compartment where food is partially digested; (3) the *small intestine* consisting of three sections – duodenum, jejunum and ileum – where digestion and absorption primarily occur, and finally (4) the *large intestine* consisting of cecum and colon where the gut flora ferments, breaks down the more complex structures in our food, such as dietary fibre, into more absorbable form, and also metabolises other bioactives. Figure 9.1 gives a diagrammatic illustration of GIT showing only the salient parts. While the food moves from the mouth to the stomach through the esophagus (old spelling: oesophagus), then into the small and large intestines before the rejected parts leave the tract through the rectum and anus, the liver, gall bladder and the pancreas play a critical role in aiding digestion by providing enzymes and the right chemical environment.

   The movement of food through the GI tract, after it leaves the mouth, is largely caused by *peristalsis*. In the mouth, it is chewing followed by swallowing – which are by and large voluntary movements. Once swallowing occurs, the flow of the food through the GI tract is involuntary. Peristalsis is essentially initiated by a local narrowing and expanding movement of a muscle – which creates a disturbance that propagates down the length of the tract to create a wave which in turn pushes the food and fluid forward.

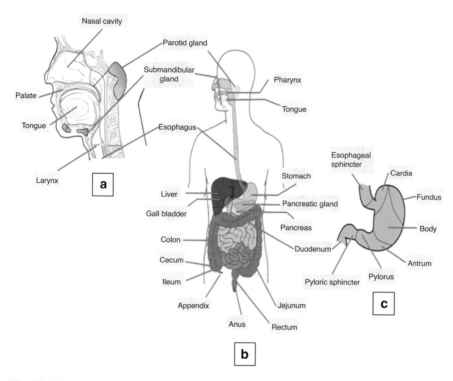

Fig. 9.1 Human gastrointestinal tract (GIT) with key parts (**a**) Oral cavity (**b**) Overall gastrointestinal tract (**c**) Stomach (After a meal, typically, the stomach can be 0.1 m at its widest point and the greater curvature length can be 0.3 m; the pyrolic sphincter diameter is around 1 cm and the stomach volume is around 0.95 litres)

9.2.1   An Introduction to the Structure and Contents of the Oral Cavity

Figure 9.1a shows the diagram of the mouth or *the oral cavity* where oral processing occurs. This organ starts at the lips and is the region up to the velum, which separates the oral cavity from the pharynx during mastication and separates the nasal cavity from the pharynx during swallowing. The epiglottis separates the esophagus from the trachea (the windpipe) and prevents food or drink from entering the respiratory tract. The size of the oral cavity varies with individuals, but on an average, the female oral cavity takes up a smaller volume of food or drink than the male oral cavity. The extent to which this happens depends on the food. Typically, the male oral cavity can hold, for instance, 20% more water in a mouthful. It is also interesting that the amount of food taken in a mouthful decreases from liquid foods to soft solids and further to hard solids; this facilitates oral food breakdown. The jaws and teeth generate extremely high forces (around 600–800 N) while chewing. The different types of teeth also perform different functions: the incisors at the front of the mouth are used for cutting; the canine teeth, which sit on each side of the incisor teeth at the front of the mouth, are used for cutting and tearing; while the premolar and molar teeth located at the back are mainly used for chewing shearing and grinding. The teeth are also sensitive to vibration while chewing, and this plays a key role in our enjoyment of crispy and crunchy foods. It is interesting to note that such tactile perception is not available to denture wearers because dentures, unlike the natural set of teeth, are not connected to the central nerve system!

The tongue is a bundle of striated muscle located at the base of the oral cavity and it plays a key role in speaking and food oral processing. It is much longer than normally perceived, beginning in the mouth with the oral part and extending almost up to the pharynx. In addition to tasting with the help of taste buds located on its surface, the tongue also plays a critical role in manipulating and swallowing food. The tongue is capable of performing a variety of muscular movements which include lengthening and shortening, thickening and thinning, curling and uncurling of its tip and edges, elevation and retraction etc., and these movements are highly coordinated by the nervous system with events like the opening and shutting of the jaws as well as amount and nature of the food in the oral cavity.

Saliva is secreted by the salivary glands in the oral cavity, and, in humans, it typically consists of 85% water and only 2% of organic and inorganic substances which include electrolytes, mucus, glyco-proteins, proteins, antibacterial compounds and enzymes. On an average, its pH is close to neutral (typically 6.7), but the pH can vary with time and when the salivary glands are stimulated. The average rate of production of saliva is about 0.5 ml min^{-1} and it can go up to 1.7 ml min^{-1} when stimulated. Almost 90% of saliva is produced by the so called major salivary exocrine glands in the mouth, although there are several hundreds of relatively minor salivary glands in the oral cavity which produce the rest. It is known that humans produce more saliva early in the morning and around noon, but it is unclear whether there is a simple direct link between, say, the texture of food in the mouth and the amount of saliva produced. At the same time, it is difficult to argue that these parameters are wholly uncorrelated. Humans are known to salivate at the smell, sight and even thought of food! Saliva plays a critical role in maintaining oral health as well as in oral processing of foods. The specific roles of the saliva include providing for: an antibacterial environment, protection of teeth, lubrication and protection, food buffering, and sensing and enhancing the mouthfeel of food.

9.2.2   Oral Processing of Foods

Oral processing essentially manipulates the food consumed to create a microstructure that generates specific textural breakdown characteristics and mouthfeel. The result of oral processing is therefore related to the food consumed, its composition and texture, and the outcome of oral processing can be used to manipulate ingredient functionality in order to create a microstructure that generates desirable mouthfeel and texture. Although oral processing senses, and indeed, modifies food texture and rheology, a mere instrumental measurement of texture and rheology of foods is not enough to tell what happens during oral processing. This is mainly because the force fields prevailing in the oral cavity and the way the food structure disintegrates in it, is very dissimilar to what happens during texture and rheological measurements. Moreover, the force fields applied not only vary with time after the food enters the oral cavity but also vary depending on the texture of the food. Thus, there is a grotesque gap between objective (i.e. instrumental) measurement of food texture and subjective sensing (i.e. in-mouth experience). Recognising this gap, Hutchings and Lillford (1988) interpreted oral processing to occur in three dimensions: the degree of structure of the food, the extent of saliva mixing and lubrication and, very critically, time. Thus, oral processing of foods is a dynamic time-dependent process involving lubrication of food for smooth flow and disintegration in the mouth in order to facilitate swallowing. The resulting mouth processing model, showing the threshold of structure disintegration and the extent of lubrication needed to enable swallowing, is illustrated in Fig. 9.2. Oral processing of solid and semisolid foods involves five interrelated destructive steps from the point of food consumption: (1) first bite, (2) chewing or mastication, (3) saliva mixing, (4) bolus formation (Latin word for a ball – i.e. food forming distinct lumps after mixing with saliva, ready to swallow) and (5) swallowing.

Although the *first bite* is a distinct step, the steps from mastication to saliva mixing and bolus formation happen concomitantly. The first bite happens immediately after food consumption and its outcome is very closely related to the textural properties of the food. In terms of timescale, it happens within seconds of food

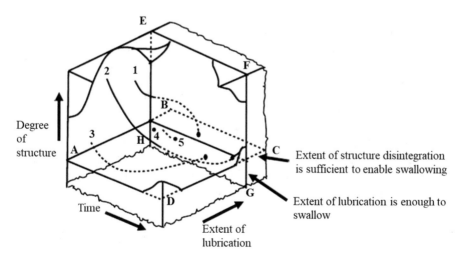

Fig. 9.2 Mouth processing model (Hutchings and Lillford 1988) showing the plane ABCD as the threshold of structure disintegration, below which swallowing can occur, and the plane EFGH as the threshold of lubrication extent, beyond which swallowing can occur. The curves consisting of solid and dotted sections – with the dotted section indicating readiness to swallow – show dynamic disintegration of (1) tender juicy stake, (2) tough dry meat, (3) dry sponge cake, (4) oyster and (5) liquid beverages

Table 9.1 Texture transformation occurring during oral processing

Stage	Upon food consumption/ first bite	Chewing and mastication	Just prior to swallowing
Key textural properties	Thickness	Creaminess	Consistency
	Firmness	Fattiness	Roughness/ smoothness
	Crunchiness	Smoothness	After-feel
	Melting	Cohesiveness	Deformability
	Brittle fracture	Deformation	Homogeneity

In addition, there is structural disintegration, bolus formation as well as astringency and after taste developments

consumption and covers a wide range of textural features, including hardness, springiness, cohesiveness, etc. The force applied in the first bite is highly dependent on the mechanical and geometrical nature of the food. Plastic materials which deform with the biting force without showing any signs of relaxation, tend to have longer first bites of up to 2 s, whereas brittle foods like chips which fracture under the biting force, take half the time. In addition to the food texture, its geometry also influences the first bite and its outcome. The biting forces for thicker foods, for instance, tend to be higher than thinner foods. Chewing or mastication fragment the food particles small enough so that they are well mixed and properly lubricated by the saliva to form a coherent bolus that can be swallowed safely and comfortably. The reduction is particle size of the food and mixing with the saliva also creates high interfacial areas for the enzymes in saliva to act. Chewing or mastication also transform the "bulk characteristics of food" such as thickness, firmness, crunchiness etc. into surface characteristics such as consistency, roughness, after-taste, astringency etc. A tabular representation of the texture transformation in the oral cavity is shown in Table 9.1. The texture and structure characterisation of each stage can be undertaken by determining: (1) rheology – the understanding of bulk deformation of products under applied shear stress, (2) tribology – the study of friction and lubrication between surfaces in relative motion, (3) microscopy – to visualise the microstructural changes of a product before and after mastication and (4) Particle size analysis which influences mouthfeel and surface properties.

By developing a detailed understanding of how products break down during oral processing and deliver textural, mouthfeel and flavour profiles provides valuable insights for product development, so that the right ingredients can be chosen or optimum processing conditions can be selected for the development of desirable sensory attributes.

The modelling of oral processing of foods has been studied extensively from different angles, and most models are either inconclusive and/or extremely complicated. Some approaches are based on the biomechanics of jaw movement, teeth crushing and swallowing (e.g. Prinz and Lucas 1997; de loubens et al. 2011; Harrison et al. 2014), while others investigate the physics of bolus formation (e.g. Witt and Stokes 2015). Then there are papers dealing with perceptions specific ingredients like salt in the oral cavity (Le Reverend et al. 2013), while a PhD thesis (Gray-Stuart 2016) provides a good review and chemical engineering modelling insights.

Problem 9.1 A piece of toast (2.7 g, 7.9 cm^3), toast with margarine spread (mass and size not significantly different)), small piece of cake (5.6 g, 9.2 cm^3), medium piece of cake (8.6 g, 14 cm^3) and large piece of cake (13.2 g, 20.0 cm^3), were each masticated separately by volunteers, to record the following data (Adapted from: Gavio and Van der Bilt 2004):

	Toast	Toast with spread	Small sized piece of cake	Medium piece of cake	Large piece of cake
Saliva flow rate (ml min^{-1})	8.64 (0.43)	7.74 (0.19)	7.97 (0.27)	7.32 (0.20)	7.42 (0.10)
Chewing time (s)	23.8 (5.8)	20.5 (4.5)	17.4 (5.2)	23.0 (6.5)	30.7 (7.2)
Number of chewing cycles until swallowing	37.6 (9.9)	32.4 (7.2)	28.4 (7.3)	36.9 (9.8)	46.4 (10.5)

Standard Deviation value given in parenthesis (SD), Means represent six replicates, $n = 6$

(i) Does the type of food chewed appears to influence the saliva flow rate in the mouth? Establish this by undertaking a suitable statistical approach. Assume that all data follow normal distribution.
(ii) Estimate the chewing cycle duration for each food and comment on the answer

(i) Assuming access to a statistical software is available, the data can be subjected to a normality test to confirm normal distribution, for example, in SPSS Kolmogorov-Smirnov or Shapiro-Wilk test.

Assuming normal distribution to be verified, the data can be subjected to one-way ANOVA to assess whether the numerical differences between means of the saliva flow rates for all the food types chewed are statistically significant. It is worth noting that such ANOVA results can only reveal whether a significant difference exists somewhere in the data set; but not specifically where.

If the ANOVA results do not verify statistically significant differences, we can assume that there was no influence of the type of chewed food on the saliva flow rate within the experimental errors.

If the ANOVA results confirm statistically significant differences, a means multiple comparison test can be applied for revealing specifically which means of saliva flow rates differ. There are several multiple comparison tests, often referred to also as Post-Hoc Tests, that can be used within a statistical software package; e.g. in SPSS: LSD, Bonferroni, Scheffe, etc.

For the practically minded food engineer with access only to a simple Excel spreadsheet, it is possible to go up to the ANOVA step, but unfortunately, a normal

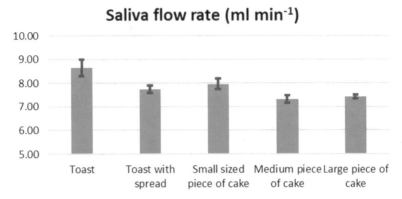

Fig. 9.3 Plot of average saliva flow rate including 95% confidence intervals for different types of foods chewed

Excel software package does not include any multiple comparison tests for means. It is however possible to estimate, for each type of food chewed, the 95% confidence interval using the excel formula:

$$= \text{CONFIDENCE.NORM}(\text{alpha}: 0.05, \text{Standard Deviation}, \text{Replicate numbers})$$

which allows the construction of a plot of the average saliva flow rate values along with 95% confidence interval bars, shown in Fig. 9.3. Observing if the 95% confidence intervals overlap or not, it is possible to state whether there are statistically significant differences between the different types of food chewed.

Key inferences
The highest Saliva flow rate was noted in the case of plain toast (i.e. without the spread). "Toast with spread" & "Small sized piece of case" shows similar saliva flow rate (since the 95% confidence intervals overlap), which is lower than in the case of the toast without the spread, but greater than the medium and larger pieces of cake, which incidentally gave similar values of saliva flow rates.

(ii) The chewing cycle duration for each food can be obtained by simply dividing the mean chewing time by the number of chewing cycles until swallowing, to give the following times for each type of food chewed, respectively: 0.63 s, 0.63 s; 0.61 s; 0.62 s; 0.66 s. Without performing any statistical analysis, it appears that the duration of the chewing cycle is the same regardless of the type of the food.

9.3 Gastric Processing of Foods

Once the bolus formed in the oral cavity is swallowed, peristalsis causes the food to reach the end of the esophagus, where a muscle, known as the lower esophageal sphincter (see Fig. 9.1c), relaxes and lets the food enter the stomach. This sphincter is normally closed in order to stop the contents of the stomach from refluxing back into the esophagus. Figure 9.1c shows a diagram of the shape and the key parts of the stomach Functionally, the stomach can be separated into two regions: the proximal stomach, comprising the cardia, fundus and body – which is thought to act as a food reservoir; and the antrum, or distal stomach, comprising the antrum and the pylorus – where ingested food remains before it moves into the distal stomach (Fig. 9.1c). The cardia is filled with mucin-secreting cells. The fundus is lined with mucosa forming thick folds, known as rugae, where parietal cells secrete HCl and chief cells secrete pepsinogen which is converted into pepsin upon contact with acid. The antrum has a smooth surface filled mainly with mucus-secreting cells. The distal stomach is thought to be the main location of the physical breakdown of food in the stomach, where antral contraction waves (ACWs), or the peristaltic contractions of the stomach wall, act to crush and grind food particles. When the action of the stomach is completed, it slowly empties its contents, known as chyme, into the small intestine through the pyloric sphincter.

The post-prandial (after food consumption) motility (or natural movement) of the stomach wall is highly critical to its digestion function. There are three main types of movement: the first type of movement starts in the proximal region as a slight indentation of the wall; the second type is a series of regular-peristaltic antral contraction waves (ACW), with a frequency of approximately 0.05 Hz, that originate at the middle of the stomach and propagate circumferentially toward the pylorus; and the third movement is the tonic contraction of the entire gastric wall, which allows the stomach to accommodate itself to varying volumes. All three movements, which are controlled by neuro-harmonal mechanisms depending on the food consumed, can be superimposed to produce highly complex mixing patterns in the stomach contents. There have been attempts to simulate the mixing patterns using computational fluid dynamics (CFD) (Ferrua and Singh 2010). Such simulations show the mixing is far more intense in the distal regions and this is expected because the cross-sectional area of this region is smaller than the proximal region. Therefore, most of the crushing and homogenisation seem to be occurring in the distal region of the stomach. Moreover, simulations of fluid flow and mixing for viscous contents in the stomach show relatively low homogenisation levels occurring. But these are simulations which assume a fixed value of stomach content viscosity for the purpose of running the CFD simulations. As a matter of fact, the stomach induces significant lowering of the viscosity of foods while mixing and shearing. For instance, Marciani (2000) experimentally found that over 40 min, the viscosity of a 0.01 Pa s solution ingested was reduced to 0.005 Pa s, and that of a 11 Pa s solution was reduced to 0.3 Pa s.

A very important task of the stomach is to empty the gastric contents into the small intestine in an orderly manner. Gastric emptying rates depend not only on the amount and properties of the food present (mainly fat and protein content), but also on the feedback control of the small intestine. It is believed that the control of the last factor dominates. The gastric emptying rate is itself not uniform: it is high initially and then slows down. If V_o is the initial volume of the stomach contents at the start of emptying and V is the volume at any time during emptying,

$$-\frac{dV}{dt} = \gamma V \qquad (9.1)$$

where γ is the rate (or decay) constant for gastric emptying. Equation (9.1) can be solved to yield the volume fraction remaining in the stomach at any time during emptying as:

$$\frac{V}{V_0} = e^{-\gamma t} \qquad (9.2)$$

The time taken for half the stomach contents to empty is known as the *half-time of emptying* – a common parameter used to describe the emptying of liquids from the stomach. Thus the half time for emptying is the value of $t = t_{1/2}$ when $V/V_0 = \frac{1}{2}$, which is related to the decay constant γ by the following equation:

$$t_{\frac{1}{2}} = \frac{\ln 2}{\gamma} \qquad (9.3)$$

This half-life has been measured experimentally for liquids consumed, and there is evidence to suggest that it is not dependent on the volume of food consumed as well as it its viscosity. The latter is not surprising because the stomach is known to lower the viscosity of the food consumed. However, the half-life for emptying appears to increase with the nutritional calorie content of the food, and typically, if the nutritional calorie content of the liquid consumed increases from 100 to 500 kcal, the half-life for emptying increases from 50 to 100 mins. Such studies using solid/semi-solid foods are not readily available. Given that significant absorption of the food takes place down-stream of the stomach, the dependence of gastric emptying times on nutritional calorific value reinforces the belief that the gastric emptying rate is controlled by intestinal feedback.

In summary, the stomach performs three main tasks: (1) it accepts and stores the swallowed food and liquid, by relaxing the muscles in the upper or proximal part; (2) it mixes the food with the digestive juices and homogenises it mainly with the help of muscles in the distal part; and finally, (3) it empties its contents slowly into the small intestine.

9.4   Processing of Foods in the Small Intestine

Most of the digestion and absorption of the food consumed occurs in the small intestine. It is estimated that almost 90% of digestion and absorption occurs here. The stomach and the large intestine contribute to the remaining digestion. The straightened length of the small intestine depends on age, with 0.2 m being the length in a new-born rising to just over 6 m in an adult. It is interesting to note that the small intestine length is believed to be roughly three times the height. This long tube is divided into three sections: duodenum (0.25 m in length), jejunum (2.5 m) and ileum (3.6 m). As shown in Fig. 9.1b, the duodenum begins just after the stomach ends, but it goes around the pancreas and returns to join the rest of the small intestine. The wall of the small intestine and colon is composed of four layers: mucosa (or mucous membrane), submucosa, muscularis, and adventitia. The submucosa secretes an alkaline mucus that neutralizes the gastric acid in the incoming chyme. The inner walls of the small intestine show mucosal folds with the number of folds being high up stream of the intestine and progressively dropping downstream to such an extent that the Ileum hardly exhibits any folds.

The muscles of the small intestine mix food with digestive juices from the pancreas, liver, and intestine, before pushing the mixture forward for further digestion. The pancreas makes a digestive juice and deliver it to the small intestine through small ducts. The digestive juice secreted by the pancreas breaks down carbohydrates, fats and proteins. Carbohydrates are broken down to simple sugars and monosaccharides like glucose. Pancreatic amylase can also break down some carbohydrates into oligosaccharides. Enzymes such as trypsin and chymotrypsin, secreted by the pancreas, act upon proteins, peptides and amino acids. Lipases secreted from the pancreas, act on fats and lipids in the diet. The lipases are helped by bile salts secreted by the liver and delivered to the small intestines through the gall bladder. The lipase is water soluble, but the fatty triglycerides are not. The bile salts hold the triglycerides in an aqueous environment until the lipase can break them into the smaller parts that can be easily absorbed.

The broken down the nutrients are absorbed by the inner walls of the small intestine into the blood stream. The nutrients are rendered small enough so that they may pass, or "be transported", across the epithelial cells. The absorption of nutrients can occur through five possible mechanisms or *driving forces*: (1) passive diffusion, (2) active transport (3) facilitated diffusion, (4) co-transport (or secondary active transport), and (5) endocytosis. *Passive diffusion* is the natural movement that follows a concentration gradient (conventional mass transfer). *Active transport* refers to movement against or up the gradient, normally induced by chemicals such as proteins in the cell membranes acting as "pumps" to draw the molecules against the gradient. *Facilitated diffusion* refers to the movement of substances from an area of higher to an area of lower concentration like passive diffusion, but by using a carrier protein in the cell membrane. *Co-transport* uses the movement of one molecule through the membrane from higher to lower concentration to power the movement of another from lower to higher. And finally, *endocytosis* refers to the cell membrane

Table 9.2 Transport mechanism for the absorption of each food category[a]

Food	Breakdown product	Transport mechanism
Carbohydrate	Glucose	Co-transport with sodium ions
Carbohydrate	Galactose	Co-transport with sodium ions
Carbohydrate	Fructose	Facilitated diffusion
Proteins	Amino acids	Co-transport with sodium ions
Lipids	Long chain fatty acids	Diffusion into intestinal cells, where they are combined with proteins to create chylomicrons
Lipids	Monoacylglycerides	Diffusion into intestinal cells, where they are combined with proteins to create chylomicrons
Lipids	Short chain fatty acids	Passive diffusion
Lipids	Glycerol	Passive diffusion
Lipids	Nucleic acid digestion products	Active diffusion via membrane carriers
Vitamins B and C		Passive diffusion
Vitamins B_{12}	In combination with an intrinsic factor from stomach	Active transport
Vitamins A, D, E and K		Absorbed with dietary fats
Minerals		
Na^+		Passive diffusion and active transport
Cl^-, I^-, NO_3^-		Passive diffusion or active transport with Na^+
Mg^{2+}, PO_4^{-3}		Active transport
Ca^{2+}	Stimulated by calcitriol (an active form of vitamin D)	Active transport

[a]Taken from: https://courses.lumenlearning.com/nemcc-ap/chapter/chemical-digestion-and-absorption-a-closer-look/

physically engulfing the molecules it requires. The transport mechanism involved in the absorption of each category of food is given in Table 9.2. In addition to the absorption of food, the small intestine also manages water consumed. On an average, c.a. 9 litres of fluid enter the small intestine of which around 2.3 litres are ingested through food and drink and the remaining are from secretions of the GI tract. Almost 90% of water is absorbed in the small intestine, which is essentially driven by the concentration gradient of water (or osmosis) since the concentration of water is higher in chyme than it is in epithelial cells. Some of the water leaving the small intestine is absorbed in the colon, while the rest exits with faeces.

Modelling flow and mass transfer phenomena occurring in the small intestine is extremely complicated given the multiphase nature of the chyme as well as the chemical reactions and other passive and active transport occurring within it. A computational fluid – structure interaction duodenum model that mimics peristaltic movements, chyme fluid flow, mixing and absorption of glucose through the porous

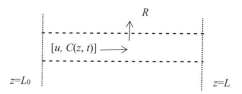

Fig 9.4 A laminar flow reactor representing the absorption of food in a section of the small intestine. The chyme from the stomach flows at a mean velocity of $\underline{u}$; the concentration of a nutrient at any axial position z at any time is $C(z,t)$; and R represents the rate of nutrient absorption per unit volume; $z = L_0$ is the axial position in the intestine where the food is ready for absorption which ends at $z = L$

duodenum wall has been developed (Hari et al. 2012). Likewise, Moxon et al. (2016) developed a model for glucose absorption in the small intestine assuming glucose absorption to follow passive diffusion. Such models are essentially *in silico* models (i.e. based on computer simulations) and have not been backed up by *in vivo* or *in vitro* experimental investigations.

Conceptually, the small intestine can be approximated to operate as a *laminar flow reactor* which is postulated to give the same radial concentration at each axial position; see Fig. 9.4. The nutrient enters and leaves the intestine through the flow of the chyme in and out of the intestine. If $\bar{u}$ is the mean convective velocity of chyme flow, the net flow of the nutrient per unit chyme volume at any axial position, also known as advection, is $-\bar{u}\frac{\partial C(z,t)}{\partial z}$. If there is concentration gradient in the axial direction, the nutrient can also enter by axial dispersion. If D_L is the axial dispersion coefficient, the net flow of nutrient by this mechanism is $D_L\frac{\partial^2 C(z,t)}{\partial z^2}$. If the rate of absorption of the nutrient is R, the rate of change of concentration at this axial position is given by (Fogler 2017):

$$\frac{\partial C(z,t)}{\partial t} = D_L\frac{\partial^2 C(z,t)}{\partial z^2} - \bar{u}\frac{\partial C(z,t)}{\partial z} - R \qquad (9.4)$$

Each of the terms in Eq. (9.4) has the units kg m^{-3} s^{-1}. The rate of absorption R depends on the nutrient transport mechanism. Suppose $C(z,t)$ represents passive diffusional transport, e.g. transportation of water or Na$^+$, then R can be expressed in terms of mass transfer coefficient, k, as $ka\,f\,C(z,t)$ where **a** is the interfacial area for mass transfer per unit chyme volume or unit volume of the cylindrical tube equal to $4/d$ where d is the diameter of the cylinder. The term f is an arbitrary multiplication factor introduced to represent the enhancement of the interfacial area due to the presence of folds along the length of the intestine. Moxon et al. (2016) have assumed f to take a value of 12 based on earlier research. It may be noted that the driving force for mass transfer has been assumed to be the axial nutrient concentration $C(z,t)$ because the nutrient absorption renders its concentration to zero at or just around the wall. Thus, Eq. (9.4) can be written as:

$$\frac{\partial C(z,t)}{\partial t} = D_L \frac{\partial^2 C(z,t)}{\partial z^2} - \bar{u}\frac{\partial C(z,t)}{\partial z} - \frac{4kf}{d}C(z,t) \tag{9.5}$$

Moxon et al. (2016) assumed an ideal plug flow reactor and ignored the term containing axial dispersion coefficient. These authors further assumed that $z = L_0$ is the axial position in the intestine where the food is ready for absorption, and $z = L$ is the axial position at which the absorption ends, so the following initial and boundary conditions could be imposed in order to solve the differential Eq. (9.5): $C(z,0) = 0$ for all z except $z = L_0$ where $C(z,0) = C_0$; and $\frac{\partial C(L_0,t)}{\partial z} = 0 = \frac{\partial C(L,t)}{\partial z} \forall t$. Typically, the diameter of the small intestine $d = 0.036$ m; the chyme velocity through the intestine $\bar{u} = 1.7 \times 10^{-4}$ m s^{-1}; and the length of the absorption section (L_0-L) depends on the nutrient (for instance, some nutrients are absorbed only in the jejunum while others may be absorbed throughout the intestinal length). In order to solve Eq. (9.5), an expression for the mass transfer coefficient, k, must also be found. Even though the mass balance (Eq. 9.4) assumes plug flow (inviscid fluid), Moxon et al. (2016) have suggested estimating k using correlations validated for laminar flow. If the density and viscosity of the chyme are assumed to be the same as water, i.e. $\rho = 1000$ kg m^{-3} and $\mu = 0.001$ Pas; the flow Reynolds number Re can be estimated to be ($d\bar{u}\rho/\mu$) ≈ 6.1, which represents laminar or viscous flow in the case of a real fluid. When the Reynolds number is as low as this value, the mass transfer coefficient can be estimated by using the following relationship between Sherwood number (kd/D), Schmidt number [$\mu/(\rho D)$] and Re, which relates the mass transfer coefficient k with the molecular diffusivity D of the nutrient (see Chap. 4):

$$\text{Sh} = 1.62\left[\text{Re Sc}\left(\frac{d}{L}\right)\right]^{\frac{1}{3}} \tag{9.6}$$

If the definitions of Sh, Re and Sc are incorporated into Eq. (9.6), the mass transfer coefficient is given by:

$$k = 1.62\left(\frac{uD^2}{Ld}\right)^{\frac{1}{3}} \tag{9.7}$$

If the nutrient is a relatively small molecule, Stoke Einstein equation or one of its modifications can be used to determine the diffusion coefficient [see Chap. 4, Eqs. (4.15, 4.16, and 4.17)]. Thus, we have a numerical value for all the parameters to solve Eq. (9.5).

Even though the analysis presented so far may appear complex to some, it is way remote from reality. For a start, the mass balance (Eq. 9.4) assumes that only absorption takes place in the intestinal section of length ΔL, and no new nutrient is formed in the section – which is unrealistic since the small intestine is constantly breaking down food consumed into absorbable forms. Secondly, the mechanism by which the nutrient is absorbed is passive diffusion – which happens only in the case of very few nutrients (see Table 9.2). There are no robust models for R in Eq. (9.4) when active and facilitated transport of nutrients occur. Equation (9.4) also assumes constant value of u, which is also unrealistic since stomach emptying is an exponentially decaying function of time (see Sect. 9.3). As a result of chyme flowrate fluctuation and the inherent intestinal coiling with peristalsis, a steady laminar flow film may not be developed, and even the mass transfer model for passive diffusion may not be valid. Regardless, the purpose of the above analysis is not to present a conclusive theory for the flow and mass transfer occurring in the small intestine, but to show that realistic engineering models need to be, and indeed can be, developed in the future – which will inevitably be complex.

Problem 9.2 Consider absorption of a nutrient in a section of the small intestine, 2 m in length and 0.036 m in diameter, where the rate of nutrient absorption is so low that there is hardly any concentration gradient in the section. Therefore, advection can be neglected. The nutrient absorption across the intestinal wall may be assumed to occur under laminar flow conditions with passive diffusion being the operating mechanism. If the nutrient concentration is 0.005 kg m^{-3}, the flow velocity is 1.7×10^{-4} m s^{-1}, and the molecular diffusivity of the nutrient is 1.1×10^{-9} m^2 s^{-1}, how much nutrient will be absorbed in the time the chyme takes to flow through the intestinal section?

Answer The mean time taken for the chyme to travel the section is $L/\bar{u} = 2/(1.7 \times 10^{-4}) = 11{,}764$ s. The rate of absorption per unit volume, according to Eq. (9.5) is equal to $\frac{4kf}{d} C_0$. The total amount of nutrient absorbed will therefore be $\left[\left(\frac{4kf}{d} C_0 \right) \left(\frac{\pi}{4} d^2 L \right) \left(\frac{L}{\bar{u}} \right) \right]$. With $\bar{u} = 1.7 \times 10^{-4}$ m s^{-1}, $d = 0.036$ m, $L = 2$ m, and $D = 1.1 \times 10^{-9}$ m^2 s^{-1}, Eq. (9.7) yields $k = 2.3 \times 10^{-7}$ m s^{-1}. Assuming $f = 12$ as discussed earlier, the amount of nutrient absorbed is 3.67×10^{-5} kg.

9.5   Food Processing in the Large Intestine or Colon

Colon and large intestine refer to the same organ. The last part of the small intestine known as Ileum (see Sect. 9.3) connects to the cecum – the first part of the colon, which is followed by the ascending colon, the transverse colon and the descending colon (see Fig. 9.1b). The descending colon rises a bit (this section is called sigmoid colon, not marked in Fig. 9.1b) before it connects to the rectum and finally the anus. The colon removes water, electrolytes, and some other nutrients, and also forms stool. Muscles line the colon wall which squeeze the contents forward. Most importantly, the colon has a very rich microbial eco-flora attached to its wall, unlike the other upstream parts of the gastrointestinal tract, where the microbial population, if any, is nominal. Typically, the colon has over 400 bacterial species at a count of around 10^{11} per gram of stool, which represents a very complex system co-existing to define and characterise the state of health of an individual. These bacteria possess very diverse metabolic capabilities, and the enzymes secreted can act on the substrates in the colon. The enzymes produced by intestinal bacteria are important in the metabolism of several vitamins. The intestinal microflora synthesizes vitamin K, which is a necessary cofactor in the production of prothrombin and other blood clotting factors. Intestinal bacteria also synthesize biotin, vitamin B_{12}, folic acid, and thiamine; and convert indigestible carbohydrates (e.g. fibre) into short chain fatty acids. Even plant cell wall polysaccharides such as pectin, cellulose, and hemicellulose are fermented into fatty acids which can be a major source of energy for the host bacteria. The intestinal bacteria also serve to suppress new organisms, especially pathogens, that enter the system through the consumption of contaminated food or water.

Spratt et al. (2005) have modelled the colon as a tubular bioreactor in which the fluid exhibits laminar flow and the main chemical reaction occurring is the fermentation of indigestible starch into fatty acids. These authors further assume that the tubular reactor operates under steady state conditions. Thus, the partial derivative with respect to time in Eq. (9.4) vanishes and the partial derivatives can be replaced by ordinary derivatives in one variable i.e. z. Further, for simplicity, $C(z,t)$ can be replaced by C, and R, the rate of reaction per unit volume, may be assumed to be given by Monod kinetics [see Chap. 5, Eq. (5.24)] to give:

$$R = \frac{\mu_{max}C}{K_S + C}x = \frac{\mu_{max}C}{K_S + C}Y_{XS}(C_0 - C) \qquad (9.8)$$

where the biomass concentration $x = Y_{XS}(C_0 - C)$ and Y_{XS} is the biomass yield coefficient with respect to the substrate [see Chap. 5, Sect. 5.5, Eq. (5.29)]. In assuming Monod kinetics as represented by Eq. (9.8), the problem has been considerably oversimplified by assuming that the cells can be represented by a single microbial culture, whereas the colon has a highly interacting mixed culture, for which robust growth models do not exist. Thus, Eq. (9.4) may be written as:

$$D_L\frac{d^2C}{dz^2} - \bar{u}\frac{dC}{dz} - \frac{\mu_{max}C}{K_S + C}Y_{XS}(C_0 - C) = 0 \qquad (9.9)$$

Eqn (9.9) is a second order ordinary differential equation and requires two boundary conditions. Like the boundary conditions to solve Eq. (9.4), the boundary condition at $z = L$ will be $\frac{dC}{dz} = 0$. However, at $z = 0$, the concentration will not be C_0, but there will be an additional axial dispersive flux:

$$C = C_0 + \frac{D_L}{u}\frac{dC}{dz} \qquad (9.10)$$

Since the flow is assumed to be laminar, the axial dispersion coefficient, D_L, also known as the Taylor dispersion coefficient, is related to the molecular diffusivity of the substrate, D, by the well-known relationship (Taylor 1953):

$$D_L = \frac{(du)^2}{192D} \qquad (9.11)$$

The variables in Eq. 9.9 can be normalised by defining the variables as follows: $y = C/C_0$, $w = z/L$, $N_0 = L\mu_{max}/u$, $K = K_s/C_0$, and $Pe = uL/D_L$ to give:

$$\frac{1}{Pe}\frac{d^2y}{dw^2} - \frac{dy}{dw} - \frac{N_o(1-y)y}{K+y}Y_{XS} = 0 \qquad (9.12)$$

with the boundary conditions:

$$\text{at } w = 0, y = \left(1 + \frac{1}{Pe}\frac{dy}{dw}\right); \text{and at } w = 1, \ \frac{dy}{dw} = 0 \qquad (9.12a)$$

The term N_0 which can be written as $[\mu_{max}/(u/L)]$ represents the ratio of the maximum specific growth rate to the rate of flow, whereas Pe, the Peclet number, represents the extent of axial mixing caused by the strong radial dependence of the axial velocity which is a characteristic feature of laminar flow. A low value of Pe represents high levels of axial back mixing whereas higher values tend towards plug flow (characterised by uniform axial velocity profile across the entire radius). The solution of Eq. (9.12) will yield the concentration of substrate as a function of the colonic length traversed. This represents an oversimplified model of human colon as a bioreactor. The stability of any continuous bioreactor will be compromised if the flow through it is so rapid that the cells do not have enough time to grow and get washed out consequently. This has been mentioned in Chap. 5, Sect. 5.6.1; see Eq. (5.37). The washout conditions for a continuous tubular bioreactor with axial dispersion has been analysed by Fan et al. (1970) who showed that the N_0 value at washout, N_{0w}, for relatively high values of Peclet number, is given by:

$$N_{0w} = 0.25 \, Pe(1 + K) \qquad (9.13)$$

Spratt et al. (2005) estimated the Peclet number to be of the order of 1 for carbohydrate substrate at the start of the colon, which falls further downstream. Thus, N_{0w} values for washout to occur are low, which also implies that the rate of flow through the intestine must be high for washout to occur. The normal rates of flow are indeed significantly lower than those at which washout occurs, and this accounts for the high natural microbial stability of the colon. The above analysis does not even consider the biofilm attached to the wall of the colon – which can only give it added stability. It is also interesting that the low values of Peclet number estimated by Spratt et a (2005), i.e. ≤ 1, suggests that the extent of axial mixing in the colon is fairly high, and a mathematical modeller must not get carried away by its tubular appearance to conclude that the flow is closer to plug flow.

   This chapter clearly illustrates that *in silico, in vitro* and *in vivo* studies, which back one another, are needed to further our understanding of flow mixing, mass transfer and biochemical reactions occurring in the gastrointestinal tract. The subject area is still in its infancy.

References

Bornhorst GM, Singh RP (2014) Gastric digestion in vivo and in vitro: how the structural aspects of food influence the digestion process. Annu Rev Food Sci Technol 5:111–132. https://doi.org/10.1146/annurev-food-030713-092346

de Loubens C, Magninc A, Doyennettea M (2011) A biomechanical model of swallowing for understanding the influence of saliva and food bolus viscosity on flavour release. J Theor Biol 280:180–188. https://doi.org/10.1016/j.jtbi.2011.04.016

Fan LT, Erickson LE, Shah S, Tsai BI (1970) Effect of mixing on the washout and steady-state performance of continuous cultures. Biotechnol Bioeng 12:1019–1068

Ferrua MJ, Singh RP (2010) Modeling the fluid dynamics in a human stomach to gain insight of food digestion. J Food Sci 75(7):R151–R162. https://doi.org/10.1111/j.1750-3841.2010.01748.x

Fogler SH (2017) Models for non-ideal reactors. http://umich.edu/~elements/5e/18chap/Fogler_Web_Ch18_final.pdf

Gavio MBD, Van der Bilt A (2004) Salivary secretion and chewing: stimulatory effects from artificial and natural foods. J Appl Oral Sci 12(2):159–163. https://doi.org/10.1590/S1678-77572004000200015

Gray-Stuart EM (2016) Modelling food breakdown and bolus formation during mastication. PhD thesis Massey University, New Zealand

Hari B, Bakalis S, Fryer PJ (2012) Computational modelling and simulation of the human duodenum. Excerpt from the Proceedings of the 2012 COMSOL Conference in Milan. https://www.comsol.com/paper/download/151975/hari_paper.pdf

Harrison SM, Eyres G, Cleary PW, Sinnott MD, Delahunty C, Lundin L (2014) Computational modelling of food oral breakdown using smoothed particle hydrodynamics. J Texture Stud 45:97–109. https://doi.org/10.1111/jtxs.12062

Hutchings JB, Lillford PJ (1988) The perception of food texture – the philosophy of breakdown path. J Texture Stud 19:103–115

Le Reverend BJD, Norton I, Bakalis S (2013) Modelling the human response to saltiness. Food Funct 4:880–888. https://doi.org/10.1039/C3FO30106K

Marciani L (2000) Gastric response to increased meal viscosity assessed by echo-planar magnetic resonance imaging in humans. J Nutr 30(1):122–127. https://doi.org/10.1093/jn/130.1.122

Moxon TE, Gouseti O, Bakalis S (2016) In silico modelling of mass transfer & absorption in the human gut. J Food Eng 176:110–120. https://doi.org/10.1016/j.jfoodeng.2015.10.019

Prinz JF, Lucas PW (1997) An optimization model for mastication and swallowing in mammals. Proc R Soc Lond B (1997) 264:1715–1721

Rein MJ, Renouf M, Cruz-Hernandez C, Actis-Goretta L, Thakkar SK, da Silva Pinto M (2013) Bioavailability of bioactive food compounds: a challenging journey to bioefficacy. Br J Clin Pharmacol 75(3):588–602. https://doi.org/10.1111/j.1365-2125.2012.04425.x

Spratt P, Nicollela C, Pyle DL (2005) An engineering model of the human colon. Food Bioprod Process 83(C2):147–157. https://doi.org/10.1205/fbp.04396

Taylor GI (1953) Dispersion of soluble matter in solvent flowing slowly through a tube. Proc Roy Soc A219:186–203

Witt T, Stokes JR (2015) Physics of food structure breakdown and bolus formation during oral processing of hard and soft solids (2015), COFS. https://doi.org/10.1016/j.cofs.2015.06.011

Further Reading

Chen J (2009) Food oral processing – a review. Food Hydrocoll 23:1–25. https://doi.org/10.1016/j.foodhyd.2007.11.013

https://www.news-medical.net/health/What-Does-the-Small-Intestine-Do.aspx

https://courses.lumenlearning.com/nemcc-ap/chapter/chemical-digestion-and-absorption-a-closer-look/

Wang X, Chen J (2017) Food oral processing: recent developments and challenges. Curr Opin Colloid Interface Sci 28:22–30. https://doi.org/10.1016/j.cocis.2017.01.001

Chapter 10
A Selection of Engineering Methodologies for Food Product Realisation

Aim The aim of this chapter is to enable an engineer to translate product briefs, which may be provided by a marketing team or any other agency, into engineering projects. This chapter particularly focuses on the steps involved in developing flowsheets for process and plant design, methodologies used to examine the economic viability of a project, and selected methods used in project management. It is necessary to note that a variety of commercial software packages are available to execute the methodologies described in this chapter, which are widely used in industrial practice. The purpose of this chapter is to explain and illustrate the basic concepts which are used to develop such software packages.

10.1 Introduction

A key expectation of a food engineer is to be able to translate a product brief into an *environmentally sustainable* and *economically viable* manufacturing design, which results in the product being produced in required quantities whilst meeting all its keeping and eating qualities as well as its health attributes. A food engineer therefore needs to know the various stages involved, from product concept, through to manufacturing design, financial justification and analysis of process performance, in order to evolve a generic methodology for *food product realisation*. Every time a product brief is to be realised in practice, it requires an engineer to take a *systems approach* and view it as an exercise in *project management*, an inspirational -albeit somewhat dated – book on this subject being written by Obeng (1994). Even though food product realisation requires the synthesis of knowledge from all aspects of food science, this chapter will discuss four specific aspects relevant to scaled-up manufacture: (1) Process and Plant design, (2) Financial investment criteria, and (3) Project management tools. It is necessary to acknowledge here that the ideas presented in this chapter were first developed by Leach (2012) in the context of a course delivered on *Operations Management* at the University of Reading. Before

© Springer Nature Switzerland AG 2022
K. Niranjan, *Engineering Principles for Food Process and Product Realization*,
Food Engineering Series, https://doi.org/10.1007/978-3-031-07570-4_10

considering how a product brief is translated, it is important to understand what constitutes *new product development* (NPD) in the context of food business, which invariably forms the basis of a product brief.

In its broadest sense, *the development and introduction of a product not previously marketed or manufactured by a company*, or *the presentation of an existing product into a new market*, constitute NPD. Thus, NPD does not necessarily mean a completely new and creative product which no market has ever seen! More specifically, NPD can include (1) *line extensions* (e.g. canned peaches varying in variety or amount of sugar or presented in sliced or un-sliced forms or varying in cost), (2) *repositioning of existing products* (e.g. antacid being repositioned as a useful source as calcium or a breakfast cereal being positioned as a health functional food), (3) *new forms of existing products* (e.g. introduction of spreadable butter by a butter manufacturer or introduction of bite-sized portion of an already marketed chocolate bar), (4) *reformulated products* (e.g. lower calorie or lower salt and sugar version of an existing snack brand) and (5) *new packaging of existing product* (e.g. change from plastic to a lower environmental legacy packaging). All these NPD possibilities can generate the product brief – which, more often than not, is likely to be brief and lacking in detail.

10.2   Process and Plant Design

In dealing with plant design, most classical textbooks tend to assume that the engineer starts with a clean sheet of paper. It is very rare that an engineer gets to design a plant from first principles. Process and plant design is a team exercise which involves in-house engineers as well as external contractors. It is therefore necessary to appreciate the management of the design process. The preferred order of steps for approaching plant design is:

1. Develop a statement of requirement (SoR)
2. Product specifications
3. Process description
4. Services and utilities
5. Plant/building

The *Statement of Requirements (SoR)* describes the project in such a way that it not only serves to inform all relevant members of the organisation but also key external stakeholders such as suppliers and contractors. The SoR therefore informs the business itself, as well as others, about what it plans to achieve in relation to the project, be it modifying a production line or changing a production line altogether or constructing a whole new factory. Although each organisation may have its own template, it must contain the following information (Leach 2012): (1) Introduction and purpose of the project, (2) Schedule of data and reference documents, (3) Operation requirements, (4) Financial Analysis, (5) Project team members, (6) Outline program and timescales, (7) Drawings of the production line or the factory area and

(8) Proposed tender format. Schedule of data and reference documents (Item 2) must include as much data as possible such as quality and quantities of the new product to be made (with perhaps a photograph of the product), recipe, ingredients needed along with their specifications, cost and margin basis for the product at various manufacturing volumes, processing steps and stages including timings, hygiene and cleaning requirements, hazard analysis and critical control point (HACCP) (Wallace et al. 2011) and packaging requirements. Operation requirements (Item 3) must include raw material, intermediate and final product handling characteristics, environmental issues including noise and odours, any fume extraction, machinery requirements, machinery performance criteria, process effectiveness and yields. Financial analysis (Item 4) can be based on Net Present Value (NPV) or Internal Rate of Return (IRR). It must be recognised that all the information needed to prepare a SoR may not be available, especially at the very start, but it is necessary to make it as detailed as possible. It is an evolving document and may require several iterations to complete, but attention must be paid to every detail, which will ensure that the statement is not merely an information document, but an effective plan for the project to remain on time and within cost.

Product specification is not merely stating which product is to be made, but it must contain all product attributes, the type of food (ready to eat food or requiring cooking), ingredients used and risks they pose, storage and cooking instructions, HACCP analysis of ingredients, processing steps and finished product. It is also necessary to describe the intended market for the product, packaging to be used and how it is distributed and consumed. A key information is the volume of the product to the sold, projected sales profile, and method of transportation to the market. In other words, the product description must be comprehensive document which can be used to design the plant and processes for estimating costs as realistically as possible.

Process description becomes much easier once the product and HACCP have been defined, because the individual manufacturing steps and control parameters can be specified. Normally, process description occurs in two stages: an outline stage followed by a detailed step by step process. The outline process only describes the stages of manufacture; it does not show volumes or wastes or specific process details. *Process flow maps or flow charts* can be constructed using software that are readily available. Figure 10.1 shows an illustration of an outline flow map. It is necessary to recognise that, during processing, the ingredients get partially transformed in each stage to finally result in the end-product. For instance, if tomatoes are being processed, the raw tomatoes are inspected, washed and stored. Once peeling commences, the process changes from *low risk operations* to *high care operations*; the latter operations are shown within the dashed rectangle in Fig. 10.1. The outline flow chart can then be expanded to obtain a more detailed flow chart which shows more detailed process steps and process streams, and their interrelationships. For example, if in Fig. 10.1, the raw materials included vegetables, salt/seasoning and meat/fish, there could be three separate stores – an ambient store for salt etc., a chilled store for vegetables and a freezer for meat and fish. High care operations would commence when some of these ingredients are brought together and the whole flowsheet can evolve depending on the detailed processing steps such

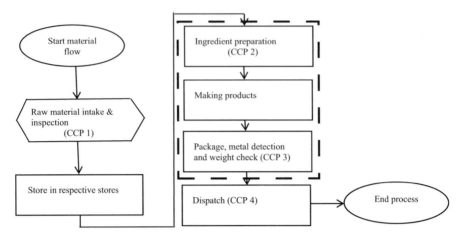

Fig. 10.1 Example of an outline process flow map. (Adapted from Leach (2012)). In general, the start and end are always included in the chart and placed within oval shape; a process step is indicated within a rectangle; decision point is indicated within diamond; and the arrow indicates the direction of flow. Critical control points (CCP) are located at any step where hazards can be either prevented, eliminated, or reduced to acceptable levels. Typical CCPs identified are also indicated in a flow diagram. The operations enclosed within the dashed rectangle are high care operations. The ingredients are irreversibly transformed from the point they enter this area

as mixing, heating/cooking, cooling, packaging etc. The flow chart evolved will consist of the functional blocks for processing as well as indicate the functional area. But the actual space needed and how the processing rooms may be designed and constructed, can only be estimated after the product volumes, manufacturing rates, physical size of the processing equipment and machinery, utilities needed (steam, compressed air, hot and cold water etc), volumes of wastes generated, plant requirements in terms of air conditioning, plantroom requirements etc. are determined. The chart therefore continues to develop progressively, with more and more details added to it. It must be mentioned that this is also an iterative process and earlier versions of the flow chart can change and evolve as more details are learnt about the process.

Problem 10.1 Develop a flow chart to manufacture a new *chilled organic herb and tomato soup*, 5 m³/day, SKU 250 ml cartons (SKU is the Stock Keeping Unit usually in the form of a scannable bar code printed on product packaging. The label allows vendors to automatically track the movement of inventory).

An outline flow diagram is shown below in Fig. 10.2. At the left end of the flowsheet, the *ingredients* are mentioned (tomatoes, herbs, water and sundries such as salt). The weighing in the case of solid ingredients and flow meter in the case of water, help to *dispense* the ingredients. Peeling and chopping represent *preparation*, which is followed by processing operations such as *mixing, pasteurisation, chilling* and *packaging*, prior to *dispatch*.

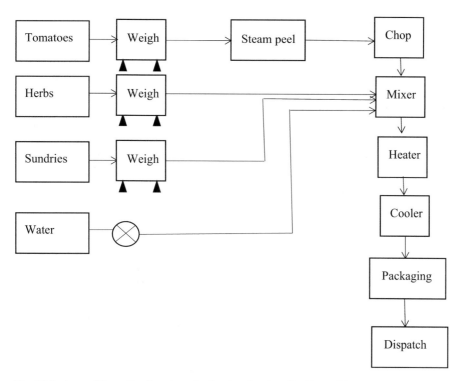

Fig. 10.2 A possible outline flowsheet for the situation described in Problem 10.1

This is an open-ended problem and one can add more details to this flowsheet. For example, if the recipe specifies that the proportion of tomatoes to herbs to sundries to water is 10:1:1:3 and the process operates continuously, the mass flow rates of all ingredients can be estimated in order to produce 5 m³/day of soup. Calculations can be undertaken to estimate the power consumed by the mixer depending on the viscosity and other properties of the blend, the power or steam required for pasteurisation which will depend on the time-temperature combination employed, the utility required for chilling etc. One can also consider modifications to the flowsheet. The simplest modification could be the decision of the business to produce two SKUs, say, 250 and 500 ml cartons. It is also possible to disperse the herbs, sundries and water and introduce this blend into the mixer to blend with the tomatoes. The flow sheet will change with process modifications, and the ingredient and utility throughputs could also change if the process changes. It is worth reflecting on the various possibilities and how each possibility will change the flowsheet.

10.3   Financial Investment Criteria

Although process economics is a mature subject with several books written on it (e.g. Maroulis and Saravacos 2008), the purpose of this section is not to cover all aspects of process economics, but specifically focus on investment criteria and economic viability of a project, be it for the improvement of a process line or the introduction of a new food product. The investment criterion is based on a basic principle that the accomplishment of the project will yield a return on the investment made within a given time frame. Whether this return is worth the endeavour or not, is entirely subjective. Some investors may be looking to maximise financial returns (and most do!), while charities may consider it worthwhile to invest in a project even if it means taking a financial loss, so long as the good causes for which the charities were set up in the first place are accomplished. The criteria for investment may be expressed in different ways, which will be considered below. But all these criteria incorporate a very basic principle – *the time value of money*, i.e. the value of a Pound (GBP) or a US Dollar (USD) today is worth more than its value sometime in the future. This is because, at any point in time between today and a future date, there is an opportunity to invest this money and make a return on the investment. The three commonly used investment criteria, which represent returns on investment, involve (1) estimating returns on investment (RoI) (2) the application of a discounted cash flow (DCF) model and (3) the application of a model based on the internal rate of return (IRR).

A simple *Return on Investment* is calculated by dividing the initial investment by the anticipated annual profit to be generated. Thus:

$$\text{RoI (years)} = \frac{\text{Value of the investment}}{\text{Anticipated annual profit}} \qquad (10.1)$$

The above model assumes that there will be a profit each year where the annual income exceeds the annual expenditure. RoI is commonly used when a quick "back of the envelope" assessment is needed to be made on the potential of an investment. At best, RoI can inform on how long it takes for the invested money to be returned. It is not suitable for comparing projects of dissimilar investments and durations. It also does not allow for testing of various scenarios.

The *Net Present Value (NPV)* method is often used to estimate the likely profitability of an investment or a project. In this method cash flows (income, expense, profit etc) are reduced to their present value i.e. to time zero, by assuming an interest rate i. If I is the value of the income and E that of expense in any year, the profit that year is $P = (I - E)$. If the project lasts n years, the *NPV* can be estimated as:

$$NPV = (I - E)_0 + \frac{(I - E)_1}{1 + i} + \frac{(I - E)_2}{(1 + i)^2} + \ldots + \frac{(I - E)_n}{(1 + i)^n} \qquad (10.2)$$

It is necessary to note that the interest rate i has been assumed to be constant, and the formula can be corrected if the interest rates vary year on year. In terms of profit P, Eq. (10.2) can simply be written as:

$$NPV = \sum_{j=0}^{n} \frac{P_j}{(1 + i)^j} \qquad (10.3)$$

It is also worth noting that inflation reduces the present value further. Note that an amount X, n years into the future is worth $[X/(1 + f)^n]$ now, if f is assumed to be the uniform rate of inflation. As an investment criterion, $NPV > 0$ establishes economic viability. If two alternative situations are being considered, an investor would probably select the alternative with a higher value of NPV. It may be noted that NPV can be used to assess a revenue dominated cash flow as above, or a cost dominated cash flow. If there are two equipment options available to an organisation with different capital and running costs, the NPV for both options can be estimated and the one with a lower NPV may be selected.

Discounted cash flow (*DCF*) is a valuation method used to estimate the value of an investment or a project based on its future cash flows. If CF_1, CF_2, $\ldots$, CF_n are the projected cashflows, the *DCF finds their present values* by using a *discount rate*, r, as follows:

$$DCF = \frac{CF_1}{1 + r} + \frac{CF_2}{(1 + r)^2} + \ldots + \frac{CF_n}{(1 + r)^n} \qquad (10.4)$$

Note that the equation for *DCF* is like Eq. (10.2) because both equations estimate the present value of future cash flows. In order for an investment to be financially worthwhile, the *DCF* must be greater than the value of the investment. The reliability of this comparison obviously depends on the value of the discount rate r, which an investor has to estimate, as well as the future projections of the cash flows. It is not desirable to use *DCF* if future cash flows cannot be accessed with reasonable accuracy.

Future value (*FV*) can also be used to compare investments, so that the alternative with the maximum future worth (in case of net revenue) or with the minimum future worth (in case of net cost) can be selected for implementation. If P is the initial investment and $R_1, R_2, R_3, \ldots, R_n$ are the net revenues at the end of each year, the *FV* at the end of n years at a rate of interest i is simply given by:

$$FV = -P(1 + i)^n + R_1(1 + i)^{n-1} + R_2(1 + i)^{n-2} + \ldots + R_n \qquad (10.5)$$

The future and present values are obviously related as follows:

$$FV = PV(1 + i)^n \qquad (10.6)$$

Internal rate of return (IRR) is yet another metric used to examine the viability of any investment, and it is the interest or discount rate that makes the *NPV* of all cash flows equal to zero. In other words, it is the value of i in Eq. (10.2) which makes $NPV = 0$. If *IRR* is greater than the target rate (which may be the interest rate set by a bank, for instance), then it is worthwhile proceeding with the investment. In case two alternative investments are being compared, the investment with greater IRR will be preferred.

The investment criteria explained above may appear somewhat abstract, but Problems 10.2, 10.3, 10.4, and 10.5 solved below will illustrate the applications of the methodologies described above.

A glossary of terms commonly used in the economic appraisal of a project is given in Table 10.1. A typical cumulative cash flow of a project is shown in Fig. 10.3. The monetary flows (possibly after correction for the time value of money) are plotted against time. The initial stages of the project involve investment (and only expenditure), say, in buying land, constructing the premises and the plant (line AB in Fig. 10.3). The cumulative cash flow is negative at this stage. Once the plant is commissioned, the product is produced, and the revenues start flowing in. With time, the cumulative cash flow goes upwards, breaks even (Point C, Fig. 10.3) and eventually becomes positive. It is worth reflecting what happens after E is reached in Fig. 10.3. At some point in time the process cessation may occur when the project ends. It is possible that after the point E, and before process cessation, the profit may remain steady or even drop off due to increased mainte- nance cost of an ageing manufacturing plant.

Problem 10.2 A food company planning to expand its production capacity wishes to assess three technology options with different initial outlays and annual revenues summarised in Table 10.2 All three technology options may be considered to have a life span of 10 years and the interest rate is 6%. Suggest the best technology option.

Solution For   Option

$$1 \; NPV = -1.2 \times 10^6 + 4 \times 10^5 \left[\frac{1}{1+0.06} + \frac{1}{(1+0.06)^2} + \cdots + \frac{1}{(1+0.06)^{10}} \right] = £1.74 \times 10^6$$

Likewise, for Option 2, $NPV = £2.42 \times 10^6$; and for Option 3, $NPV = £1.88 \times 10^6$. Thus, the *NPV* for technology option 2 seems to be the highest and therefore it appears to be the best.

Problem 10.3 A company producing frozen food is planning to buy a fully automated freezer. If the freezer is purchased under down payment, it costs £1.6 M. The company also has an alternative payment option which involves payment of 15% of the cost at the time of purchase followed by 10 annual payments of £200,000 each. Which is the better alternative for the company if the annual interest rate is 6%.

Table 10.1 A glossary of terms commonly used in the economic appraisal of projects

Capital	**Money borrowed that is paid back over a certain number of financial periods (usually years) at a set amount each year.**
Revenue	Income and Expenditure in a business which is wholly accounted for within the same financial period.
Negative cash flow (NCF)	Monetary outflows or expenditure (e.g. the money spent to buy things).
Positive cash flow (PCF)	The money flowing into a business from sales. Do not confuse with profitability as this is the point that a business makes money after paying off the capital borrowed and the interest on that borrowing
Debt	Money borrowed that must be paid back.
Interest	This is the fee to be paid on the debt or money borrowed. In business terms, the cash flow to do this is often referred to as servicing the debt.
Discount rate	An interest rate within the context of a *present value*, *net present value* (*NPV*) or *discounted cash flow* (*DCF*) calculation.
Payback period	It is the period (time) required to pay back the capital borrowed plus the accrued interest.
Time value of money	It is the value of a sum of money together with interest earned over a given period
Present value (*PV*)	It is the value on a given date of a future payment (or a series of future payments) discounted to reflect the time value of money.
Net present value (*NPV*) or Net present worth	It is the sum of the present values of the individual cash flows both incoming and outgoing (e.g. sales and expenditure)
Future value (*FV*)	It is the nominal future sum of money that a given sum of money will be worth at a specified time in the future at a given interest rate
Discounted cash flow (*DCF*)	It is an analytical method used to value a project or asset using the time value of money.
Return on investment (*RoI*)	A simple payback period calculation indicating how much money needs to be paid back each period (usually each year) on the capital borrowed. It usually does not take account of the time value of money.
Internal rate of return (*IRR*)	A rate of return used in capital budgeting to measure and compare the profitability of proposed investments. It is particularly useful in comparing different investments which have different durations and capital requirements. It is also the effective interest rate to the given point in the future that the proposed investment moves to profitability
Profitability	The point at which the *NPV* curve passes to positive cash flow
Yield/Margin	The difference between manufacturing price and sale price (a negative margin is a loss)

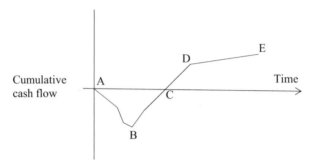

Fig. 10.3 Typical cumulative cashflow of a project. AB represents the investment stage. B represents the maximum borrowing depth, but it also represents the point where revenues commence to flow and pay back begins. C is the payback or breakeven point after which the cashflow is positive. CDE represents the period of profit (Leach 2012)

Table 10.2 Data for Problem 10.2

Technology option	Initial outlay ($£ \times 10^{-6}$)	Annual revenue ($£ \times 10^{-5}$)
Option 1	1.2	4.0
Option 2	2.0	6.0
Option 3	1.8	5.0

Table 10.3 Data for Problem 10.4

| | Initial investment | Net revenue | | | | | |
		Year 1	Year 2	Year 3	Year 4	Year 5	Year 6
Product A	500,000	50,000	100,000	150,000	200,000	250,000	300,000
Product B	700,000	70,000	140,000	210,000	280,000	350,000	420,000

Solution Since the company has to pay 10% of the cost of the freezer initially, i.e. £160,000, the net present worth of the alternative payment is given by:

$$NPV = 160,000 + 200,000 \left[\frac{1}{1+0.06} + \frac{1}{(1+0.06)^2} + \cdots + \frac{1}{(1+0.06)^{10}} \right]$$

$$= £1.63 \text{ M}$$

It therefore appears that there is no significant difference between the two options based on *NPV*. The decision whether to go for down payment or the alternative can therefore be based on other criteria. For instance, the company may find it more convenient to go for the alternative option. It would be desirable to reflect on different scenarios under which the company can go for each option.

Problem 10.4 A food company has two options: to manufacture Product A or Product B. The initial investment and the net returns for each year over a period of 6 years are tabulated below in Table 10.3. Based on the figures tabulated, which product must the company manufacture if the interest rate is assumed to be 6%?

Solution The decision on which product to manufacture could be based on the future value of the project at the end of year 6.
    For Product A, using Eq. (10.5):

$$FV = -500,000(1.06)^6 + 50,000(1.06)^5 + 100,000(1.06)^4 + 150,000(1.06)^3$$
$$+ \ldots + 300,000$$
$$= £452,271$$

Likewise, the FV for Product B can be shown to be: £916,884. The future worth of manufacturing Product B seems more than Product A which decides in favour of manufacturing Product B.

Problem 10.5 The initial outlay of a 5-year project is **£800,000** and the net annual revenue is projected to be **£250,000**. Estimate the internal rate of return (*IRR*).

Solution If i is the interest rate, based on Eq. (10.2), the *NPV* is given by:

$$NPV = -800,000 + 250,000 \left[\frac{1}{1+i} + \frac{1}{(1+i)^2} + \ldots + \frac{1}{(1+i)^5} \right]$$

The *IRR* is the value of i which will result when *NPV* is set equal to 0 in the above equation. It is, however, essential to note that the above equation is non-linear, and a numerical or a graphical method will be necessary to estimate the value of i. A simple way to estimate i is to use Excel spreadsheet to characterise the variation of *NPV* with i, It is clear from Fig. 10.4 that the *IRR* is approximately 0.17 or 17%.

10.4 Project Management and Monitoring Tools

A *project* can be any activity which generally involves more than one individual, and where *time* and *cost* needs to be controlled. Examples of typical projects in food business are new product development (NPD), factory change, health and safety inspection, expanding the capacity of an existing production line etc. *Project management* defines a methodology to accomplish the project, helps coordinate local and remote activities, provides control and tracking mechanisms, and enables the objectives of the project to be achieved in a stipulated time and within budget. The four main steps in project management involve: (1) Defining the project and its

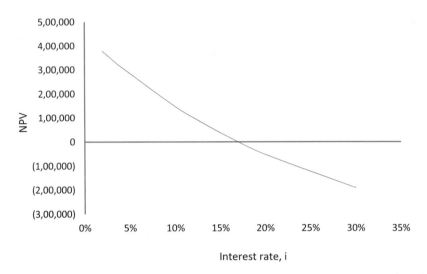

Fig. 10.4 Plot of Net Present Value (NPV) against interest rate for the data given in Problem 10.5

scope, (2) Planning the project, (3) Monitoring or tracking the project, and (4) Closing out (Obeng 1994). A project must be closed out, and if it is not, it is more likely to be a hobby!

The planning phase of a project involves (a) dividing the project into distinct activities or tasks, (b) estimating time requirement for each task, (c) establishing precedence relationships among the tasks and (d) constructing and analysing network diagrams representing the tasks from start to finish. Estimating the time requirement for each task and scheduling the tasks based on precedence requirements is critical from the point of view of monitoring the project. The precedence relations provide information on the tasks which precede and follow a given task as well as other tasks which are concurrent with it. In addition to time, the resource allocation for each task is also critical especially for monitoring whether a project is operating within budget or not.

Network diagrams show the different tasks as well as the interrelationships between the various tasks. The *Activity on Arrows* (AOA) diagram is commonly used in project management for this purpose. In an AOA diagram, arrows are used to represent activities; and nodes represent activity dependencies. Any activity coming into a node is a predecessor to any activity leaving the node; this is illustrated in Fig. 10.5. Further common nomenclatures are illustrated in Fig. 10.6. The basic rules for constructing an AOA network diagram are: (1) The network must have a unique starting node and completion node. (2) No activity has more than one arc in the network. (3) No two distinct activities will have the same starting node and the same ending node.

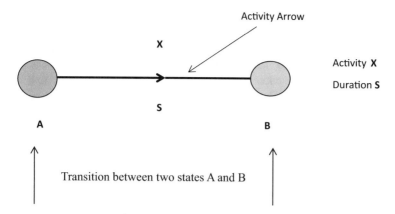

Fig. 10.5 Basic notation used in activity on arrow (AOA) diagrams

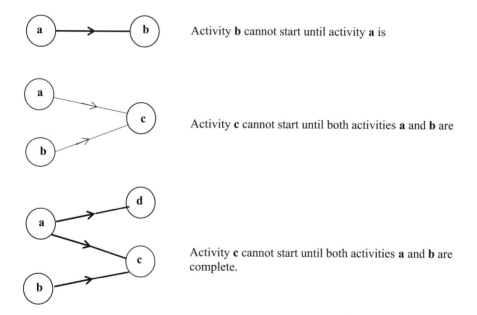

Fig. 10.6 Further notations used in activity on arrow (AOA) diagrams

10.4.1 *Critical Path Method (CPM) and Gantt Charts*

This method uses network diagrams to estimate project durations and the best routes for organising the tasks. CPM is ideally suited for projects with many activities or tasks, where on-time completion for each stage is critical. The network diagrams and time calculations show critical path and sub tasks. A very important point to note is that the critical path is the *longest path in the network diagram from start to finish,* and it determines the duration of the project. The concept and methodology of CPM is best illustrated by solving Problems 10.6 and 10.7.

But before this is done, it would be desirable to note that the network calculations enable us to produce time charts or Gantt Charts (named after Henry Gantt) which gives a clear calendar schedule of the whole project, including the starting time and ending time of each task. The Gantt chart also shows the task interrelationships and gives a good visual description of the project timescales. Gantt chart is also used for resource levelling purposes. When resources such as manpower, money, equipment etc. are limited, the Gantt chart also helps to adjust the schedule of activities, especially *the non-critical activities* (i.e. activities not appearing on the critical path of a network diagram). The method used to deduce a Gantt chart is better explained through Problem 10.6.

Problem 10.6 A team consisting of food engineers and marketing personnel are asked to prepare a quick feasibility report to manufacture a variant of a protein enriched drink which a company is already manufacturing. The activities involved are tabulated below. Draw the network diagram and undertake critical path calculations. Also produce a Gantt chart giving a typical calendar schedule of all activities (Table 10.4).

Solution Figure 10.7 shows the network diagram for the above activities. Since activities A and B do not have prerequisites, the project could start with either activity. All other activities have prerequisites and the network diagram shows the dependencies as well as the time duration of each activity in days. There are three pathways from start to finish:

Table 10.4   Data for Problem 10.6

Activity	Description	Immediate predecessors	Duration (days)
A	Forecasting sales volumes		5
B	Competition analysis		7
C	Select ingredients and modify current process design	A	5
D	Prepare production schedule	C	3
E	Estimate production cost	D	2
F	Decide on the sales price	B, E	1
G	Undertake viability study and prepare the final report	E, F	7

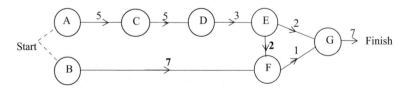

Fig. 10.7 CPM for the situation described in Problem 10.6

1. Start-A-C-D-E-F-G-Finish which will take 22 days
2. Start-A-C-D-E-F-G-Finish which will take 23 days
3. Start-B-F-G-Finish which will take 15 days.

The longest duration pathway is (2) taking 23 days which is the critical path; this path will enable the completion of all tasks given above including task B which is the only activity not on the critical path. The activities on the critical path have *no slack* and must be started on a fixed day counting either from the start or the finish. For example, the project must start with activity A; activity B starts after 5 days, activity C after 10 days and so on. To clarify further, activity G must start after 16 days regardless of whether we count the number of days from the start or the number of days backwards from the 23 days it takes to finish the project. But activity B – which is not on the critical path – has a *slack*. It could either start along with A – which is the *earliest start* day; or counting backwards from the finish, it could start 8 days after the start, which represents the *latest start*. In fact, activity B can be started any time in between, and the project will still conclude after 23 days. Thus, the *slack time* or *float* for activity B, which is the difference between the earliest and latest start times, is 8 days.

The Gantt chart can now be developed using Excel based software for which templates are available. A simple Gantt chart for the tasks shown in Problem 10.6 is given below (Fig. 10.8).

Problem 10.7   The Table 10.5 lists the stages of a project to be completed with estimates of the corresponding times needed. For each task, the cost of each week of operation is £5000 which includes the site operational and establishment costs.

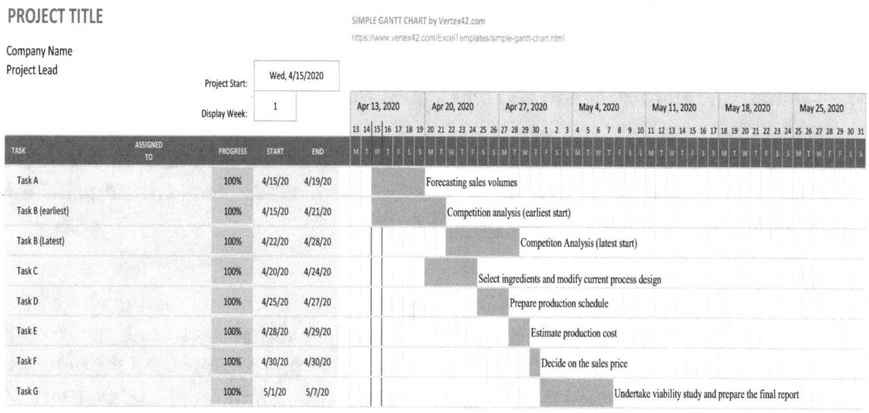

Fig. 10.8 Gantt chart for the various tasks listed in Problem 10.6 with 15 April 2020 as the starting date

Table 10.5 Data for Problem 10.7

Task	Immediate predecessors	Duration (weeks)	Maximum possible time savings (weeks)	Cost of each week saved[a] (£ or GBP)
A	–	4	1	6000
B	–	5	2	6000
C	–	4	1	3000
D	A & B	4	0	0
E	B & C	5	2	5000
F	D	4	1	5000
G	E	6	3	3000
H	F & G	3	1	7000

[a]The cost of saving a week on the activity is the additional resources needed for accelerating the task, which may be by employing additional human resource or better equipment/machinery

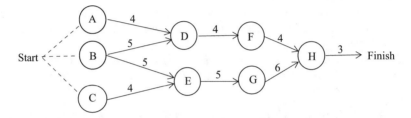

Fig. 10.9 Network diagram for the situation described in Problem 10.7

(a) Draw a network diagram and identify all possible routes to project completion, stating the project duration in each case.
(b) Identify the critical path and estimate the project cost.
(c) Tabulate the Earliest Start (ES) and Latest Start (LS) and Slack Time (ST) for each task
(d) Based on the number of weeks which can be saved for each task, how can the critical path obtained in part (b) be reduced, so that four new critical paths are created. Also estimate the new project cost.

Solution
(a) Figure 10.9 shows the network diagram for the given activities. Since activities A, B and C do not have prerequisites, the project could start with any one of these activities. All other activities have prerequisites and the network diagram shows the dependencies as well as the time duration of each activity in weeks.

   The possible routes from start to finish are:

Start–A-D-F-H-finish = 15 weeks
Start-B-D-F-H-finish = 16 weeks

Table 10.6 Tabulated values of earliest and latest start of activities given in Problem 10.7 (see Table 10.5)

Task	Earliest start (week number)	Latest start (week number)	Slack time (weeks)
A	0	1	1
B	0	0	0
C	0	1	1
D	5	8	3
E	5	5	0
F	8	12	4
G	10	10	0
H	16	16	0

Start-B-E-G-H-finish $= 19$ weeks

Start-C-E-G-H-finish $= 18$ weeks

(b) The critical path, which takes the longest duration of 19 weeks, is {start, B, E, G, H, finish}. Each of the tasks from A to H costs £5000 per week. Regardless of which path is chosen, an equivalent of 35 weeks (obtained by adding the numbers under the column entitled "Duration") will be required to accomplish all tasks from A to H. The total project cost is $= 35 \times 5000 = £175{,}000$.

(c) The method to find the earliest start and latest start is already illustrated in Problem 10.6 and the results can be tabulated as follows (Table 10.6)

(d) The critical path takes 19 weeks and the next longest duration path (Start-C-E-G-H-finish) takes 18 weeks. In order to reduce the critical path duration, and make two critical paths, 2 weeks must be taken away from the critical path. As the tasks are given in terms of *cost to save a week*, it is best to start with the least expensive task to save, and work onto the more expensive tasks later. If the duration of task B is reduced by 1 week (in order to bring in next highest path duration), the saving for reduction in the project duration will be £5000 but the additional cost of reducing task B by a week will be £6000. Hence on-cost is £1000 and duration is 18 weeks. Thus, the project duration can be reduced by 1 week by investing an additional resource of £1000 which will raise the project cost to £176,000. Although late in the program, the next reduction can be on task G by 3 weeks which will cost £9000 but save £15,000 (because the project duration is now 15 weeks), giving a net saving £4000. The project cost will therefore become £172,000, which also makes four critical paths of 15 weeks each. (It may be noted that such cost and time adjustments can, in general, be done in different ways and there need not be a unique way of accomplishing more critical paths based on a network).

10.4.2 Project Evaluation and Review Technique (PERT)

The CPM analysis in Sect. 10.4.1 assumes that the time duration required to accomplish each task is certain, precise, and deterministic. This is far from true, and estimations of time durations tend to be more probabilistic. PERT uses three estimates for the time duration of each task or activity: a is an *optimistic time* estimate of a task which can be accomplished if it is executed very well; b is a pessimistic time estimate if the execution is very bad; and m is an estimate of the most likely time of execution. The probabilistic data for project activities are assumed to follow *beta distribution* – which represents a family of continuous probability distributions (Gupta 2011). The formula for mean (μ) and variance (σ^2) of the three-point beta distribution are given by:

$$\mu = \frac{a + 4m + b}{6} \tag{10.7}$$

$$\sigma^2 = \left[\frac{b - a}{6}\right]^2 \tag{10.8}$$

If i represents a critical task or activity, the project duration will be $\sum \mu_i$ with a variance of $\sum \sigma_i^2$. It is also possible to estimate the probability of completing a project within a stipulated time period, say C. If x is the actual project duration, the beta distribution can be approximated to a *standard normal distribution* whose statistic is given by:

$$z = \frac{x - \mu}{\sigma} \tag{10.9}$$

Problem 10.8 Consider the activities given in Problem 10.7 with the same set of "immediate predecessors" (Table 10.7). The optimistic (a), pessimistic (b) and likely (m) estimates of task durations are tabulated below. Estimate the expected duration and variance of each activity and the expected project

Table 10.7 Data for Problem 10.8

Task	Duration (weeks)			Expected duration (weeks)	Variance
	a	m	b	(μ)	(σ^2)
A	2	4	8	4.33	1.00
B	3	5	10	5.50	1.36
C	2	4	6	4.00	0.44
D	1	4	6	3.83	0.69
E	2	5	8	5.00	1.00
F	2	4	6	4.00	0.44
G	3	6	9	6.00	1.00
H	2	3	5	3.17	0.25

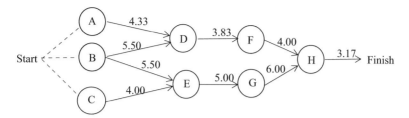

Fig. 10.10 Network diagram for the situation described in Problem 10.7 with expected task durations

completion time. **What is the probability of completing the project in (i) 21 weeks and (ii) 23 weeks?**

Solution The expected duration of each task and variance can be calculated using Eqs. (10.7) and (10.8), respectively. It may be noted that the network diagram in Fig. 10.10 is the same as shown in Fig. 10.8, except that, for convenience, the times indicated for each task are the *expected duration*. The critical path is (Start-B-E-G-H-Finish) and the expected time to complete the project is 19.67 weeks with the total variance of the tasks on the critical path being $\sigma^2 = (1.36 + 1.00 + 1.00 + 0.25) = 3.61$, whence $\sigma = 1.9$ weeks.

(i) The probability of completing the project in 21 weeks is estimated by determining the z value according to Eq. (10.9) as $(21 - 19.67)/1.9 = 0.7$. For this z value, the area under a standard normal distribution curve to the left of the mean is 0.758. (This value can be found out from any statistical table or internet resource). Thus, the probability of completing the project in 21 weeks is 75.8%.

(ii) When $x = 23$ weeks, $z = 1.75$ and the area under the normal distribution curve is 0.96, which implies that there is a 96% chance of project completion in 23 weeks.

References

Gupta AK (2011) Beta distribution. In: Lovric M (ed) International encyclopedia of statistical science. Springer, Berlin, Heidelberg

Leach KG (2012) Process realisation. In: Brennan JG, Grandison AS (eds) Food processing handbook, vol 2. Wiley-VCH, pp 623–666. ISBN: 978-3-527-32468-2

Maroulis ZB, Saravacos GD (2008) Food plant economics. Taylor and Francis. ISBN 13: 978-0-8493-4021-5

Obeng E (1994) All change! The project leader's secret handbook. Pearson Education Ltd, London. ISBN 10: 0273622218 ISBN 13: 9780273622215

Wallace CA, Sperber WH, Mortimore SE (2011) Food safety for the 21st century: managing HACCP and food safety throughout the global food chain. Wiley Blackwell. ISBN 978-1-4051-8911-8

Problems

1. Two mixtures of ethanol and water are contained in separate tanks. The first mixture contains 40 wt% ethanol and the second contains 70 wt% ethanol. If 200 kg of the first mixture is combined with 150 kg of the second, calculate the mass and composition of the product?

    (Ans: mass of product stream 350 kg and its ethanol composition is 25.9%)

2. A fizzy drink is made by dissolving 1 litre of carbon dioxide at $0\ °C$ and 3 atmospheres pressure in a litre of flavoured sucrose syrup (sucrose concentration is 10% by mass and flavour concentration is negligible). What is the mass and mole fraction of carbon dioxide in the fizzy drink? The density of the syrup is $1330\ \text{kg m}^{-3}$; ideal gas constant $R = 8.314\ \text{kJ kmol}^{-1}\ \text{K}^{-1}$; $1\ \text{atm} = 101{,}330\ \text{Pa}$; molecular weight of sucrose $= 342$; molecular weight of water $= 18$ and molecular weight of $CO_2 = 44$).

    (Ans: Mass fraction of $CO_2 = 0.0044$ and its mole fraction $= 0.0059$)

3. Air at $60\ °C$ and 10% relative humidity (RH) is passed over a bed of diced potatoes at the rate of 20 kg dry air per second. If the rate of evaporation of water from the potatoes, measured by the rate of change in weight, is $0.16\ \text{kg s}^{-1}$, estimate the temperature and RH of the air leaving the dryer. You may use the psychrometric chart.

    (Ans: temperature of the air leaving the dryer is $41\ °C$ and its RH 42%)

4. The figure below shows a drier for removing water from a flow of solids, which operates under adiabatic conditions. Fresh air is available at $15\ °C$ and relative humidity 50%. The air enters the drier at $50\ °C$. The air leaving the drier is at $34\ °C$ and has a dew point of $31.5\ °C$.

© Springer Nature Switzerland AG 2022
K. Niranjan, *Engineering Principles for Food Process and Product Realization,*
Food Engineering Series, https://doi.org/10.1007/978-3-031-07570-4

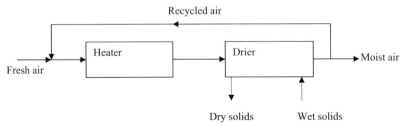

Dry solids   Wet solids

Using a psychrometric chart, determine: (i) the humidity of air entering the drier; (ii) the fraction of air leaving the drier which has been recycled; (iii) the mass of water removed from the solids per kg of dry fresh air used; and iv) the temperature of air entering the heater and the heat required per kg of water evaporated. If necessary, assume specific heat of air $= 1007$ J kg^{-1} K^{-1} and specific heat of water $= 2000$ J kg^{-1} K^{-1}.

(Ans: (i) 0.022 kg /kg dry air; (ii) 0.31; (iii) 0.01 kg/kg dry air; (iv) Air enters at 21 °C and heat required per kg water evaporated is 4402 kJ)

5. Moringa Oleifera seeds have the following composition: moisture 6.43%, crude protein 43.26%, fat 21.36%, fibre and ash 16.53% and carbohydrate 12.42%. If 1000 kg of seeds are subjected to the process shown in the flowsheet below, estimate the compositions of the various streams if (a) 35% of the total oil can be expressed by crushing, (b) the mass ratio of feed to solvent, i.e. hexane, in the extractor is 1:1, and 95% of the solvent employed is recovered, (c) 75% of the protein present is precipitated into stream D. A number of assumptions have been made in terms of the composition of the various streams. Critically discuss these and suggest more realistic compositions which one is likely to observe in practice.
   (Ans: Stream A 925.24 kg, Stream G 74.76 kg oil, Stream C 786.4 kg, Stream B 138.84 kg, Stream S 45.98 kg, Stream D 324.45 kg, Stream E 461.95 kg, Stream P 441.83 kg and Stream W = 20.12 kg)

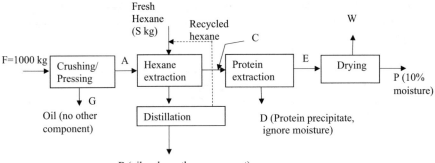

6. Imagine you are a process technologist working for the "Piste Urp Breweries", where beer is to be pumped at a rate of 10 m^3 per hour to a vessel 25 m above a reference level through a 25 mm internal diameter stainless steel pipe system. This system consists of 30 m run of straight pipe and two elbows with friction equivalent to 20 pipe diameters each. The viscosity of the beer is 1.2 mPas, and its

density is 1075 kg m^{-3}. A representative of "SuperEff Pumps" considers these details and recommends that you buy Model Super-Eta 05, which is fitted with a 5 kW motor. Explain in detail, how you will examine whether this is a sensible suggestion. Friction factor in straight pipes can be estimated by the following equations: $f = 16/\text{Re}$ for laminar flow and $f = 0.079 \text{ Re}^{-0.25}$ for turbulent flow when the frictional pressure drop is given by Eq. 2.17.

(Hint: Based on the data, estimate the frictional and gravitational head and the required power rating of the pump, following which the appropriateness of the suggestion can be examined)

7. It is required to transfer a concentrated puree (density $\rho = 1100 \text{ kg m}^{-3}$, Newtonian viscosity, $\mu = 850 \text{ mPas}$) from a storage vessel to a mixer, at a rate of $15 \text{ m}^3 \text{ h}^{-1}$ under laminar flow conditions. The length of the pipe work is 80 m; and the safe working pressure of the pipe, which will be supplied by *The M-n-M Pipe Company*, is 5 bar. The Company can supply the necessary pipe in diameters ranging from 1 inch to 8 inches, in increments of 0.5 inches. Select the smallest pipe which can be used to deliver the puree to the mixer.

   Data: 1 bar $= 101$ kPa and 1 inch $= 0.025$ m. The frictional pressure drop is given by Eq. 2.17 and the friction factor (f) for laminar flow is related to the pipe Reynolds number (Re) by the equation: $f = 16/\text{Re}$, where $\text{Re} = (Du\rho/\mu)$; note that D is the internal diameter of the pipe and u is the mean fluid velocity.

8. The terminal settling velocity of a particle in a fluid is given by $u_t = \sqrt{\dfrac{2V_p g\left(\rho_p - \rho_f\right)}{C_d A_p \rho_f}}$

   (see Eq. 2.23 and explanation given in Sect. 2.7 for notations). In a processing plant, spherical bean particles of different sizes are introduced horizontally into a sauce flowing upwards through a pipe at a velocity 0.017 m/s. Determine the diameter of the smallest particle that would begin to settle, given: the density of the bean particle is 1100 kg m^{-3}; the density of the sauce is 950 kg m^{-3}; the viscosity of the sauce is 500 mPas; the gravitational acceleration is 9.8 m s^{-2}; and the drag coefficient $C_d = 24/\text{Re}_p$ where Re_p is the flow Reynolds number around the particle, given by $dpu_t\rho_f/\mu_f$.

   (Ans: A neutrally buoyant particle will have a diameter of 0.01 m, and any particle greater in diameter than this one will tend to settle).

9. The following pressure gradients were obtained with a fruit puree formulation (density $= 1310 \text{ kg m}^3$), when the puree was made to flow through a series of pipes at different flow rates:

Pipe diameter (m)	Volumetric flow rate $(\text{m}^3 \text{ s}^{-1}) \times 10^5$	Pressure gradient $(\text{Pa m}^{-1}) \times 10^{-5}$
0.010	7.83	0.340
0.010	15.66	0.480
0.025	49.17	0.086
0.025	98.34	0.121
0.030	70.00	0.065
0.030	140.00	0.092

Based on the tabulated values of the three parameters and assuming power law flow $\tau = k\gamma^n$ through the pipes, estimate the values of k and n. What is the pressure drop if it is required to transfer the puree through a 0.035 m diameter pipe, 20 m long, at a flow rate of 5 m^3 h^{-1}?

(Ans: $k = 2.67$ Pas^n; $n = 0.5$; pressure drop $= 1.24 \times 10^5$ Pa).

10. A commercial cold storage facility is to be erected to maintain an internal temperature of -18 °C with a maximum surrounding air temperature of 25 °C. The walls are to be constructed of concrete blocks 20 cm thick followed by 15 cm of bioplastic foam on the inside. The external convective heat transfer coefficient is 10 W m^{-2} K^{-1} and the internal convective heat transfer coefficient is 6 W m^{-2} K^{-1}. The external dimensions of the storage facility are $40 \times 20 \times 7$ m high. Determine the maximum refrigeration load due to heat gains from outside air, given that the thermal conductivities of concrete and foam are 0.75 and 0.05 W m^{-1} K^{-1}, respectively. Also assume that ceiling and floor loss rates per m^2 are one-half of those for the walls.

(Ans: Maximum refrigeration load is approximately 20 kW)

11. Sucrose syrup is to be heated from 50 °C to 70 °C by making it flow in a pipe having an internal diameter (d) of 0.023 m at the rate of 40 litres per minute. The heating can be achieved by steam flowing outside the pipe in such a way that the inside wall of the pipe always remains at 80 °C. Calculate the average heat transfer coefficient inside the pipe and estimate the length of the pipe required to heat the syrup given that: the mean density of the syrup (ρ) is 1200 kg m^{-3}; the mean specific heat of the syrup (c_p) is 3120 J kg^{-1} K^{-1}; the mean thermal conductivity of the syrup (k) is 0.46 W m^{-1} K^{-1}; the mean viscosity of the syrup in the bulk of the pipe (μ) is 3.8 mPas while that at the wall (μ_w) is 2.3 mPas; and the heat transfer coefficient for forced convection in a pipe, at a flow velocity u, is given by: Nu $= 0.023$ Re$^{0.8}$ Pr$^{0.33}$ $(\mu/\mu_w)^{0.14}$

where Nu $= (hd/k)$; Re $= (du\rho/\mu)$; Pr $= (c_p\mu/k)$. You may make assumptions if necessary, but please state them clearly.

(Ans: Average heat transfer coefficient in the pipe is 2582 Wm^{-2} K^{-1}, and the required pipe length is 14.7 m. It is worth pondering how this length can be practically provided).

12. A volatile flavour condenses on the shell side of a shell and tube heat exchanger, consisting of 60 thin-walled horizontal tubes of diameter (d) 20 mm and length 2 m, through which water flows in parallel at a total flow rate of 20 kg/s. If condensation takes place at 80 °C and water enters the tubes at 20 °C, evaluate: (i) the overall heat transfer coefficient; (ii) the temperature of water at the tube exit; and (iii) the rate of condensation of the flavour. The external (i.e. shell side) heat transfer coefficient is 4 kWm^{-2} K^{-1} and the internal heat transfer coefficient, h_i, can be deduced from: Nu = 0.023 Re$^{0.8}$ Pr$^{0.4}$, where, in usual notations, Nu = $h_i d/k$, Re = $du\rho/\mu$, and Pr = $c_p \mu/k$. For water assume: $\mu = 0.8 \times 10^{-3}$ Pas, $\rho = 1000$ kg/m^3, $c_p = 4100$ J kg^{-1} K^{-1} and $k = 0.6$ W m^{-1} K^{-1}. The latent heat of condensation of flavour is 510 kJ kg^{-1}.

(**Ans: the overall heat transfer coefficient is 2167.97 Wm^{-2} K^{-1}; the temperature of the water when it exits the tubes is 30.84 °C; and the vapour condensation rate is 1.74 kgs^{-1}**).

13. spherical oxygen bubble formed in an infinite pool of oxygen-free water transfers oxygen isothermally at a temperature T (K), without any change in its volume V (m^3). If k_m is the mass transfer coefficient (m s^{-1}) and H^* is the Henry's solubility constant for oxygen (Pa m^3 kg^{-1}), show that the rate of change of pressure in the bubble P (Pa) is given by:

$$-\frac{dP}{dt} = 6^{2/3} \frac{k_m RT}{H^* M} \left(\frac{\pi}{V}\right)^{1/3} P$$

where R is the ideal gas constant (J kmol^{-1} K^{-1}) and M is the molecular weight of oxygen. If the initial bubble pressure is P_0, show that the pressure at any time t is approximately given by: $P = P_0 \exp\left[-6^{2/3} \frac{k_m RT}{H^* M} \left(\frac{\pi}{V}\right)^{1/3} t\right]$. If Sherwood number Sh = $k_m d/D = 2$ where d is the bubble diameter and D is the diffusivity of oxygen in water = 2×10^{-9} m^2 s^{-1}, estimate how long it will take for the pressure in a 2 mm diameter bubble to fall to half its initial value. The following values may be assumed for the constants: Molecular weight of oxygen $M = 32$, $T = 298$ K, $R = 8300$ J kmol^{-1} K^{-1}, Henry's constant for oxygen $H^* = 1.2 \times 10^4$ Pa m^3 kg^{-1}.

(**Ans: 17.9 s**)

14. A yeast is being grown aerobically in batch culture on a glucose feed with concentration 60 kg m^{-3}; the growth yield coefficient is 0.5 kg yeast (kg glucose)$^{-1}$ and the maximum specific growth rate μ_m is 0.25 h^{-1}. The maximum value of the volumetric oxygen transfer coefficient $k_L a$ is 0.125 s^{-1}. It may be assumed that growth kinetics are always oxygen controlled with the specific growth rate μ being given by:

$$\mu = \frac{\mu_m C_0}{0.001 + C_0}$$

where C_0 is the concentration of dissolved oxygen. If the intended that the dissolved oxygen concentration in the medium is 0.005 kg m^{-3}, would the oxygen transfer system be adequate (a) when the biomass concentration $X = 5$ kg m^{-3} and (b) towards the end of the fermentation? (c) Estimate the dissolved oxygen concentration and specific growth rate, if towards the end of the fermentation, the oxygen transfer rate is such that it just matches the required oxygen consumption rate. The saturation oxygen concentration in the medium $C^* = 0.01$ kg m^{-3}, and the yield coefficient of the yeast against oxygen is $Y = 5$ kg biomass (kg oxygen)$^{-1}$.

(Ans: (a) and (b) The oxygen transfer rate exceeds the rate of oxygen consumption; and (c) Towards the end of fermentation, the dissolved oxygen concentration is 0.007 kg m^{-3} and the biomass growth rate is 0.0018 kg m^{-3} s^{-1}).

15. Growth of a food-grade yeast on an n-alkane feed can be represented by:

$$C_nH_{2n} + aO_2 + bNH_3 \rightarrow cCH_{1.66}O_{0.2}N_{0.27} + dCO_2 + eH_2O$$

   (a) Experiments on a feed with average composition $C_{10}H_{20}$ show that the cell growth yield on the substrate $Yxs = 0.6$ kg kg^{-1}. Calculate: a, b, c, d, e and the yield coefficient on oxygen.
   (b) The cell biomass is to be produced continuously at a rate of 25 tonnes/day in a well stirred fermenter. Growth kinetics follow Monod kinetics (Eq. 5.23) with $\mu_{max} = 0.2$ h^{-1} and $K_s = 0.1$ kg m^{-3}, and it is proposed to operate at a dilution rate of 0.18 h^{-1}, with a feed composition of 50 kg alkane/m^3. Calculate the exit alkane and cell concentrations, the fermenter volume and the fermenter productivity.

(Ans: concentration of alkane at outlet 0.9 kg m^{-3}, cell concentration = 29.46 kg m^{-3}, fermenter volume = 196.4 m^3, and fermenter productivity = 5.30 kg m^{-3} h^{-1})

16. The initial spore load in a liquid food is estimated to be $N_i = 5 \times 10^{11}$ m^{-3} (i.e. expressed as a number per unit volume). A reduction to $N = 5 \times 10^{-3}$ spores m^{-3} is considered necessary and it is proposed to carry out the sterilisation continuously in a steriliser consisting of a heat exchanger and a holding tube. The spore death kinetics at the sterilisation temperature is given by the following equation: $R_{death} = -2.5N$ where N is the spore number concentration and R_{death} is expressed in (spore numbers m^{-3} min^{-1}).

   (a) Show that the spore number concentration N and the residence time of the liquid food in the holding tube are related by the equation: $N = N_i e^{-2.5\tau}$, and estimate the residence time needed for the reduction in spore numbers given above.
   (b) If, instead of a tubular configuration, the sterilisation was carried out in a well-mixed vessel, show that the corresponding equation relating spore number concentration to residence time is given by: $N_o = \frac{N_i}{1+2.5\tau}$. How

does this residence time compare with the residence time obtained in the case of tubular plug flow?

(Ans: (a) **13.82 min** (b) $\cong$ **4 $\times$ 10^{13} min**).

17. A food factory produces flavour bottles at a rate of P bottles a day. The variable costs per bottle produced is £ $(47.73 + 0.1\ P^{1.2})$. The total daily fixed charges are £1750, and all other expenses are constant at £ 7325 a day. If the product sells at £ 173 per bottle, determine:

(a) the daily profit at a production schedule giving the minimum cost per bottle;
(b) the daily profit at a production schedule giving maximum daily profit; and
(c) production schedule at the break-even point.

(Ans: (a) **165 bottles, (b) 198 bottles, (c) 88 bottles**)

18. A product is to be produced from sugar beet molasses in a 100 m^3 batch fermenter with a typical production cycle lasting 5 days. The output of the product is typically 15% of the fermenter mass, and the yield of the product is typically 50% of the mass of molasses taken. Further, the density of the fermentation broth is 1tonne m^{-3}. The following costs may be assumed: (i) cost of beet molasses = £65 per tonne; (ii) cost of other chemicals, energy etc. per annum = £30 per te (tonne) of beet molasses used; and (iii) labour cost = £300 per day. It takes 3 years to build the production facility, which works for 330 days each year. The capital expense in year 1 is £100,000, year 2 is £ 150,000 and year 3 is £ 200,000. It is also intended to write down the capital after 8 years from the start of the project and run it for 15 years overall.

(a) Ignoring the time value of money, calculate the production cost of the product per te, and examine the viability of the project if the product sells at £400 per te.
(b) Populate the cash flow and investment table over an 8 year duration of the project under the following headings: Investment capital (K), Operating expenditure (C), Revenue Income (R), Net Balance (P) and Cumulative Balance (ΣP).
(c) Based on the figures generated in the above table, briefly answer the following:

(i) When is the company back in positive cash flow?
(ii) What is the maximum exposure of the investment and when does it occur?

(d) If the interest on capital is 8% per annum and the running costs are assumed to be uniform for the duration of the project, estimate (i) the Net Present Value of the project (ii) the breakeven time, and (iii) the breakeven price of citric acid.
(e) Estimate the "Internal rate of return" for the above project.

(Ans: (a) **£ 380.69 per te, (c)(i) between 7 and 8 years, (c)(ii) £ 450,000 at the end of year 3 (d)(i) £477520.8 (d)(ii) 7.13 years, (d)(iii) £359.8; (e) 23%**)

Answers to Problems

1. Overall mass balance: $200 + 150 = P = 350$ kg; EtOH balance: $200 \times 0.4 + 150 \times 0.7 = Py = 350y$ Or $y = 0.259$ or 25.9% EtOH

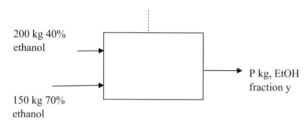

200 kg 40% ethanol

P kg, EtOH fraction y

150 kg 70% ethanol

2. Consider 1 litre of sugar syrup. Its density is 1330 kg m^{-3}. Therefore, its mass is 1.33 kg and the mass of sucrose is 0.133 kg ($0.133/342 = 3.89 \times 10^{-4}$ k mol). The mass of water is $1.33-0.133 = 1.197$ kg ($1.197/18 = 0.067$ kmol) Assuming carbon dioxide to be an ideal gas, the number of moles in 1 litre of the gas at 3 atm is given by $PV = nRT$ where $P = 3$ atm $= 3 \times 101,330$ Pa; $V = 10^{-3}$ m^3; $R = 8.314$ kJ kmol^{-1} K^{-1}; and $T = 273$ K. Thus $n = 1.34 \times 10^{-4}$ kmol $= 0.0059$ kg CO_2. Thus, the mass of fizzy drink is strictly speaking $1.33 + 0.0059$ kg $\cong 1.33$ kg.

The mole fraction of CO_2 is: $1.34 \times 10^{-4}/(0.067 + 3.89 \times 10^{-4} + 1.34 \times 10^{-4}) \cong 0.002$. The mass fraction of $CO_2 = 0.0059/(0.0059 + 1.33) = 0.0044$

3. From the psychrometric chart, Humidity of air at 60 °C and 10%Relative Humidity (RH) $= 0.013$ kg kg^{-1}. Humidity added to air in drying $= 0.16$ kg/20 kg dry air $= 0.008$ kg kg-1

Total humidity of air leaving dryer $= 0.013 + 0.008 = 0.021$ kg kg^{-1}. Following the adiabatic line on the psychrometric chart from the entry point at 60 °C and 10%RH up to the intersection of the line with a constant humidity of 0.021 kg kg^{-1}, the temperature of the air leaving the dryer is 41 °C and its RH 42%.

© Springer Nature Switzerland AG 2022
K. Niranjan, *Engineering Principles for Food Process and Product Realization*, Food Engineering Series, https://doi.org/10.1007/978-3-031-07570-4

4. Basis: 1 kg of dry fresh air. The answer to this problem extensively relies on the use of the psychrometric chart for ai-water system properties. Air temperature = 15 °C and RH = 50%. Therefore, from the psychrometric chart, the absolute humidity = 0.008 kg/kg dry air (approximately). Air leaving the drier has a dew point of 31.5C and a dry bulb temp of 34C. This yields a relative humidity of almost 90% and an absolute humidity of 0.031 kg water/kg dry air based on the psychrometric chart.

   (i) Given that the drier operates under adiabatic conditions, the air inlet to the drier will be on the adiabatic saturation line going through the point representing dry bulb temp of 34 °C and humidity of 0.032 kg/kg. Further, since the inlet temp is 50 °C, the absolute humidity of air entering the drier is approximately 0.022 kg /kg dry air.

  (ii) The heater does not affect absolute humidity. Hence the absolute humidity of air entering the drier is the same as that entering the heater = 0.022 kg/kg. If R kg of dry air is recycled, the moisture associated with it is 0.032R kg (since the air humidity is 0.032 kg water/kg dry air). Moisture with fresh air = 0.008 kg. A water balance at the mixing point, upstream of the heater, yields: 0.008 + 0.032R = 0.022, which gives R = 0.44 kg. Thus 1.44 kg of dry air leaves the drier, of which 0.44 kg is recycled. The fraction of air leaving the drier which has been recycled is 0.44/1.44 = 0.31

 (iii) Air enters the drier at 0.022 kg moisture and leaves at 0.032 kg moisture. Hence water removed from the solid is 0.01 kg/kg dry air.

 (iv) Assume a base temperature of 15 °C (i.e. that of inlet air). Enthalpy of inlet air $=0$. Enthalpy of recycle stream $= m_{air}cp_{air}(34-15) + m_{water}cp_{water}(34-15)$- $= (0.44)(1007)(19) + (0.032)(0.44)(2000)(19) = 8954$ J. If T is the temperature of air entering the drier, its enthalpy is: $(1.44)(1007)$ $(T-15) + (0.022)(1.44)(2000)(T-15) = 1513(T-15)$ J. From an enthalpy balance around the mixing point, we have: $1513(T-15) = 8954$, giving $T = 21$ C. Air therefore enters the heater at 21 °C, and its enthalpy = 8954 J. Enthalpy of the air entering the drier $= (1.44)(1007)(50-15) + 1.44(0.022)$ $(2000)(50-15) = 52{,}970$ J. Enthalpy added by the heater = 44,016 J. This evaporates 0.01 kg water. Hence heat added per kg water evaporated = 4402 kJ.

5. Basis: 1000 kg moringa oleifera seeds. Moisture 64.3 kg, crude protein 432.6 kg, fat 213.6 kg, fibre and ash 165.3 kg and carbohydrate 124.2 kg. Since 35% of oil is pressed out, G = 74.76 kg and it and it is assumed to be purely oil. Stream A = 1000–74.76 = 925.24 kg and contains the same weight of moisture, protein, fibre and ash, and carbohydrate as the feed, and the mass of fat = 138.84 kg. The fat is quantitatively extracted by the solvent, so the mass of stream C = 925.24–138.84 = 786.4 kg and it will contain 64.3 kg moisture, 432.6 kg crude protein, fibre and ash 165.3 kg and carbohydrates 124.2 kg. If the oil is quantitatively extracted into the hexane, the mass of stream B is 138.84 kg. The mass of the solid entering the extractor is A = 925.24 kg = mass of Hexane used

in the extractor. Since 95% of this hexane is recovered by distillation, the mass of hexane recovered and recycled $= 879.26$ kg. Therefore, the make up hexane needed to be added $S = 925.24 - 879.26 = 45.98$ kg. Since 75% of protein present in stream C is precipitated, the mass of protein in D is $0.75(432.6) = 324.45$ kg. Stream E will therefore weigh $(786.4 - 324.45) = 461.95$ kg, and it will contain 64.3 kg moisture, 108.15 kg protein, 124.2 kg carbohydrate and 165.3 kg fibre and ash. A mass balance around the dryer must now be performed. The overall mass balance yields $461.95 = P + W$. We can also undertake a moisture balance: $64.3 = W + 0.1P$. Solving for P and W, we get: $P = 441.83$ kg and $W = 20.12$ kg.

6. Flow rate of beer $= 10/3600 = 2.78 \times 10^{-3}$ m^3/s; velocity $= 5.662$ m/s; Re $= 126,805; f = 4.186 \times 10^3$. Effective pipe length $= 30 + (20 \times 0.025 \times 2) = 31$ m. Frictional loss $= (2 \times 4.186 \times 10^{-3} \times 31 \times 1075 \times 5.662^2)/0.025 = 357,766$ Pa. Vertical liquid head $= 25$ m $= 25 \times 1075 \times 9.8 = 263,375$ Pa. Net pressure developed by the pump $= 621,141$ Pa. The mechanical power that must be generated by the pump $= 621,141 \times 2.78 \times 10^{-3} = 1727$ W or 1.727 kW. If the pump runs at an efficiency of 50%, the electrical power consumed $= 3.4$ kW. Thus a 4 kW motor should suffice, but to allow for factors such as future capacity expansion etc. a 5 kW may be desirable. Hence the suggestion appears sensible albeit somewhat over designed!

7. The highest pressure is experienced at the upstream end of the pipe, and this can, at the most, be 5 bar. The downstream end of the pipe is discharging the puree at atmospheric pressure or 1 bar. Therefore, the pressure drop across the pipe is 4 bar or 404,000 Pa. Volumetric flow rate of puree $= 15/(3600) = 4.17 \times 10^{-3}$ m^3/s. Velocity $= 4.17 \times 10^{-3}$/area $= 5.31 \times 10^{-3} D^{-2}$ m/s. Re $= D(5.31 \times 10^{-3} D^{-2})$ $(1100)/0.85 = 6.87\ D^{-1}$. Friction factor $= 16/$Re $= 2.33\ D$. The pressure drop $\Delta P = 2f\rho u^2 L/D = 4.66D \times 1100 \times (5.31 \times 10^{-3} D^{-2})^2 (80)/D = 404,000$ (i.e. 4 bar). Hence $11.56\ D^{-4} = 404,000$ or $D = 0.0731$ m $= 2.88$ inches. The smallest pipe which can do the job safely will therefore have a diameter of 3 inches.

8. Since the vertical flow velocity is 0.017 m s^{-1}, the smallest particle that will not settle against the flow of the sauce will have a diameter d_p whose settling velocity is 0.017 m s^{-1}. The volume of the particle $V_p = (\pi/6)d_p^3$ and its cross sectional area normal to the flow is $A_p = (\pi/4)d_p^2$ giving $V_p/A_p = 0.67d_p$. Further $C_d = 24\mu_f/(d_p u_t \rho_f) = 24 \times (500 \times 10^{-3})/(0.017 \times 950)d_p^{-1} = 0.743\ d_p^{-1}$. These expressions can be substituted into the expression for u_t given, to yield:

$$0.017 = \sqrt{\frac{2(0.67d_p)(9.8)(1100 - 950)}{0.743d_p^{-1} \times 950}} \text{giving } dp = 0.01 \text{ m}$$

9. If the flow of the fruit puree were turbulent, the pressure gradient through a given pipe $(\Delta P/L) \propto u^2$ (see paragraph below Eq. 2.19), whereas $(\Delta P/L) \propto u$ if the flow were Newtonian laminar. It is clear from the table that in each of the pipes, when the velocity doubles, the pressure drop less than doubles. This is typical of pseudoplastic flows.

The general equation for pressure gradient is given by Eq. (2.15) where $f = 16/$ Re* with Re* being given by Eq. (2.20). Therefore,

$$\frac{\Delta P}{L} = \frac{2f\rho u^2}{d} = \left(\frac{32u^2}{d}\right)\left(\frac{8^{n-1}k}{d^n u^{2-n}}\right)\left(\frac{3n+1}{4n}\right)^n,$$

which can be re-written as:

$$\frac{d\Delta P}{L} = 32k8^{n-1}\left(\frac{3n+1}{4n}\right)^n\left(\frac{u}{d}\right)^n \text{ or } \ln\left(\frac{d\Delta P}{L}\right)$$

$$= \ln\left[32\,8^{n-1}\left(\frac{3n+1}{4n}\right)^n\right] + n\ln\left(\frac{u}{d}\right)^n.$$

If the tabulated values in the problem are transferred to an excel spread sheet and $\ln[d(\Delta P/L)]$ is plotted against $\ln[(u/d)]$, the gradient of the best fit line is 0.5, which is the value of n the flow index; and the intercept $\ln\left[32\,8^{n-1}k\left(\frac{3n+1}{4n}\right)^n\right] = 3.52$, which gives $k = 2.67\text{Pas}^n$. In order to find the pressure drop at a flow rate of 5 m^3 h^{-1} through a pipe of diameter 0.035 m, the flow velocity will be $(5/3600)/[(\pi/4)(0.035)^2] = 1.44$ m s^{-1}. Using Eq. (2.20), Re* = 400.55. Therefore the friction factor = 16/Re* = 0.04. The pressure gradient, given by Eq. (2.19), i.e. $\Delta P/L = 6200.3$ Pa m^{-1}, whence, $\Delta P = 124006.2$ Pa for 20 m length of the pipe.

10. The overall heat transfer coefficient across the walls, U, is approximately given by:

$$\frac{1}{U} = \frac{1}{h_o} + \frac{1}{h_i} + \left(\frac{x}{k}\right)_{\text{concrete}} + \left(\frac{x}{k}\right)_{\text{foam}} = \frac{1}{10} + \frac{1}{6} + \frac{0.2}{0.75} + \frac{0.15}{0.05}$$

which gives: $U = 0.283$ W m^{-2} K^{-1}

The above equation assumes that the walls are relatively thin in relation to the dimensions of the cold store. The rate of heat transfer from outside is: $UA\Delta T$. Here A = area of the built walls (i.e. excluding floor and ceiling). An accurate calculation of the wall area must consider the wall thickness. However, given that wall thickness is only 10% of the smallest dimension, the error entailed in basing the area on the external dimensions alone will not be too high (max. 13%). Thus, total wall area = 2(280 + 140) = 840 m^2. Hence the rate of heat transfer through the walls = 0.283 × 840 × (25 + 18) = 10,222 W, which works out to 10,222/840 = 12.17 W m^{-2}. The rate of heat transfer through the floor and ceiling is therefore: 6.05 W m^{-2}. Assuming this area to be 2 × 40 × 20 = 1600 m^{-2}, the total rate of heat transfer = 10,222 + 6.05 × 1600 = 19,958 W. The maximum refrigeration load can therefore be approximated to 20 kW.

11. The rate of hear transfer involved in heating 40 litre/min of sucrose from 50 °C to 70 °C is given by: $Q =$ (mass flow rate of sucrose) $\times$ $(c_p) \times$ (70–50). The volumetric flow rate of sucrose is 40 l/min $= 6.67 \times 10^{-4}$ m^3/s and its density is 1200 kg m^{-3}; therefore its mass flow rate is $6.67 \times 10^{-4} \times 1200 = 0.8$ kg s^{-1}, whence $Q = 0.8 \times 3120 \times 20 = 49{,}945$ W.

The sucrose flow velocity = flow rate/pipe cross sectional area $= 6.67 \times 10^{-4}$ / $[\pi/4(0.023)^2] = 1.604$ m/s. Therefore Re $= 11{,}650$. Also, Pr $= 25.77$; and $(\mu/\mu_w)^{0.14} = 1.073$. Hence from the equation given for the Nusselt number, Nu $= 129.12$, whence the inside heat transfer coefficient $h_i = 2582.34$ Wm^{-2} K^{-1} based on the definition of Nu. Since the sucrose solution is heated by steam condensing outside the pipe, the heat transfer coefficient outside the pipe h_o may be assumed to be significantly greater than the internal heat transfer coefficient, i.e. $h_o \gg h_i$. In other words, the overall heat transfer coefficient $U \cong h_i$. The rate of heat transfer $Q = UA\Delta T_{ln}$ where A is the heat transfer area $= \pi dL$, d being the pipe diameter and L - the pipe length. Also ΔT_{ln} is the log mean temperature difference, which according to Eq. (3.24) is given by:
$\Delta T_{ln} = [(80–70)–(80–50)]/\ln(10/30) = 18.2$ °C. Therefore, $Q = 49{,}945 = 2582.34A(18.2)$, whence $A = 1.06$ m^2. Therefore, $\pi\, d\, L = 1.06$, and with $d = 0.023$, $L = 14.7$ m.

12. Suppose the mass flow rate of the flavour is m kg s^{-1}, and the water leaves the tubes at a temperature of T_o. The rate of heat transfer to condense the vapour $q = m(510)$kW $= 510{,}000\, m$ W.

The rate at which water gains heat is also

$$q = 20(4100)(T_o - 20) = 82{,}000(T_o - 20) \tag{1}$$

The rate of gain of heat by water is equal to the rate at which the heat is lost by the condensing vapour, we have:

$$(T_o - 20) = 6.22m \tag{2}$$

The rate of heat transfer across the tube walls can be estimated as: $q = UA\Delta T_{ln}$. The heat transfer area A is the number of tubes multiplied by πdL, i.e. $60 \times \pi(0.02)(2)$ m$^2 = 7.54$ m^2. The log mean temperature driving force is given by: $\Delta T_{ln} = \dfrac{(80-20)-(80-T_0)}{\ln\left[\dfrac{80-20}{80-T_0}\right]} = \dfrac{T_0-20}{\ln\left[\dfrac{60}{80-T_0}\right]}$. The shell side heat transfer coefficient is 4 kWm^{-2} K^{-1} $= 4000$ Wm^{-2} K^{-1}. The tube side heat transfer coefficient h_i can be estimated from the correlation given for which the water velocity through the pipe must be determined. Given that the water flow rate through 60 tubes is 20 kg s^{-1}, it seems reasonable to assume that this flow is uniformly divided between the 60 tubes, giving a mass flow rate of 0.33 kgs^{-1} or a volumetric flow rate of 3.33×10^{-4} m^{3}s^{-1}, which can be divided by the tube cross sectional area to give a flow velocity $u = 1.06$ m s^{-1}. Thus, using the properties of water given, we have Re $= 26{,}496$ and Pr $= 5.47$. Therefore, the Nusselts number Nu $= 0.023(26496)^{0.8}(5.47)^{0.4}$ or Nu $= h_i d/k = 156.84$ or $h_i = 4705$ Wm^{-2} K^{-1}. Thus, the overall heat transfer coefficient U can be estimated from $1/U = 1/h_i + 1/h_o$ as $U = 2167.97$ Wm^{-2} K^{-1}. Thus the rate of hear transfer is given by: $q = UA\Delta T_{ln} = 2167.97(7.54)\dfrac{T_0 - 20}{\ln\left[\frac{60}{80 - T_0}\right]} =$

$16346.49\dfrac{T_0 - 20}{\ln\left[\frac{60}{80 - T_0}\right]}$. Using this result for q and Eq. (1), we have:

$$82000\,(To - 20) = 16346.49\dfrac{T_0 - 20}{\ln\left[\frac{60}{80 - T_0}\right]}$$

which yields, $T_0 = 30.84$ °C. Substituting this value in Eq. (2), we have $m = 1.74$ kgs^{-1}.

13. According to Eq. (4.19), the rate of oxygen transfer from the bubble (kg s^{-1}) is given by

$$r = k_m A C^*$$

where ΔC in Eq. (4.19) is equal to $C^*(kg\ m^{-3})$, the saturation solubility of oxygen prevailing at the interface of the bubble. This is because the oxygen concentration in the bulk liquid is given to be zero. Also, k_m is the mass transfer coefficient (m s^{-1}), A is the interfacial area of the bubble with the liquid (m^2) and $C^* = P/H^*$ where P is the oxygen pressure inside the bubble which decreases with time due to the diffusion occurring. The rate of oxygen transfer is also the rate at which the mass of the bubble (m) decreases due to mass transfer, i.e. $r = -\frac{dm}{dt}$. The mass (m) and the pressure (P) of the oxygen bubble are related by the ideal gas law: $PV = nRT = mRT/M$ where M is the molecular mass of oxygen and R is the ideal gas constant. Thus $m = \frac{VM}{RT}P$ and $r = -\frac{dm}{dt} = -\left(\frac{VM}{RT}\right)\frac{dP}{dt}$ since the volume and temperature are constants. Thus, we have an ordinary differential equation relating pressure with time as follows:

$$-\left(\frac{VM}{RT}\right)\frac{dP}{dt} = \frac{k_m A}{H^*}P$$

which can be rearranged after noting that the volume and area of the spherical bubble of diameter d are related geometrically by the equation: $A = 6^{2/3}\pi^{1/3}V^{2/3}$ (note that for a sphere, $A = \pi d^2$ and $V = \frac{\pi}{6}d^3$). Substituting the expression for A in the differential equation, we get:

$$-\frac{dP}{dt} = 6^{2/3}\frac{k_m RT}{H^* M}\left(\frac{\pi}{V}\right)^{1/3}P$$

The above first order ordinary differential equation can be solved with the initial condition, $P = P_0$ at $t = 0$, to give:

$$P = P_0 \exp\left[-6^{2/3}\frac{k_m RT}{H^* M}\left(\frac{\pi}{V}\right)^{1/3} t\right]$$

Since $Sh = k_m d/D = 2$, k_m in the above expression can be replaced by $2D/d$. Given $d = 2 \times 10^{-3}$ m and the values of the other constants, the time taken for the pressure to fall to half its initial value (i.e. $P/P_0 = 0.5$) can be calculated to be 17.9 s.

14. (a) It would be reasonable to assume that the oxygen concentration in the fermentation medium is maintained at the minimum allowable value; i.e. $C_0 = 0.005$ kg m^{-3}. Since the saturation oxygen concentration $C^* = 0.01$ kg m^{-3}, the rate of oxygen transferred or supplied from air per unit volume of the medium is $k_L a(C^* - C_0) = 0.125(0.01 - 0.005) = 6.25 \times 10^{-4}$ kg m^{-3} s^{-1}. The specific growth rate of the yeast, μ, at this dissolved oxygen concentration is $0.25(0.005)/[0.001 + 0.005] = 0.208$ kg m^{-3} h^{-1} (note that $\mu_m = 0.25$ h^{-1}) $= 5.79 \times 10^{-5}$ kg m^{-3} s^{-1}. Therefore the cell growth rate at a biomass concentration $X = 5$ kg m^{-3} is $\mu X = (5.79 \times 10^{-5})(5) = 2.89 \times 10^{-4}$ kg m^{-3} s^{-1}. Since the cell yield against oxygen is given to be 5 kg biomass (kg oxygen)$^{-1}$, the oxygen consumption rate at this cell concentration is $2.89 \times 10^{-4}/5 = 5.79 \times 10^{-5}$ kg m^{-3} s^{-1}, which is much lower than the rate at which oxygen is supplied i.e. 6.25×10^{-4} kg m^{-3} s^{-1}. Therefore, the oxygen transfer rate is adequate at a biomass concentration $X = 5$ kg m^{-3}.

(b) Towards the end of the fermentation, it can be assumed that all 60 kg m^{-3} glucose would have got converted to the biomass Since the yield of biomass against glucose is given to be 0.5 kg yeast (kg glucose)$^{-1}$, the concentration of biomass $X \cong 30$ kg m^{-3}. Since the intended dissolved oxygen concentration is 0.005 kg m^{-3}, the μ value will be the same as above, and the growth rate will be $30 \mu = 30(5.79 \times 10^{-5}) = 1.74 \times 10^{-3}$ kg m^{-3} s^{-1}. Given the yield of biomass growth against oxygen, the oxygen consumption will be $1.74 \times 10^{-3}/5 = 3.47 \times 10^{-4}$ kg m^{-3} s^{-1}, which is still lower than the oxygen supply rate of 6.25×10^{-4}. Therefore there will be no oxygen transfer limitation even towards the end of the fermentation.

(c) In this case, the oxygen supply rate is such that it just matches the oxygen consumption rate. If C_0 is the concentration of dissolved oxygen, the growth rate will approximately be $\mu X \cong \frac{0.25C_0}{0.001+C_0}\left(\frac{30}{3600}\right)$ kg m^{-3} s^{-1}. The oxygen consumption rate is therefore $\mu X/5 = \frac{0.25C_0}{0.001+C_0}\left(\frac{6}{3600}\right)$. The oxygen supply rate is $k_L a(C^* - C_0) = 0.125(0.01 - C_0)$. By equating the two rates and solving for C_0, the value of the dissolved oxygen concentration is given by $C_0 = 0.007$ kg m^{-3} (note that the equation to be solved is quadratic, so there will be two values of C_0. When the values are determined, only one root is positive – which has been reported here). The specific growth rate under this condition is $\frac{0.25(0.007)}{0.001+0.007}\left(\frac{30}{3600}\right) = 0.0018$ kg m^{-3} s^{-1}.

15. With $n = 10$, the production of yeast from the alkane may be written as:

$$C_{10}H_{20} + aO_2 + bNH_3 \rightarrow cCH_{1.66}O_{0.2}N_{0.27} + dCO_2 + eH_2O$$

(a) Balancing the number of atoms of each element on both sides of the equation gives:

$$
\begin{array}{ll}
c + d = 10 & \text{(Carbon balance)} \\
20 + 3b = 1.66c + 2e & \text{(Hydrogen balance)} \\
2a = 0.2c + 2d + e & \text{(Oxygen balance)} \\
b = 0.27c & \text{(Nitrogen balance)}
\end{array}
$$

In addition, it is given that the yield of cell growth on the carbon substrate is 0.6. The empirical formula weight of the yeast is 23.84 and that of the alkane is 140; therefore:

$$23.84c = 0.6(140)$$

We therefore have 5 unknowns and 5 equations to solve. Doing so yields: $c = 3.52$, $d = 6.48$, $b = 0.95$, $e = 8.50$ and $a = 11.08$. The yield coefficient of biomass on oxygen is: $23.84c/32a=0.24$

(b) According to Eq. 5.34, the dilution $D = \mu = \frac{\mu_{max}S}{K_S+S}$. Given $D = 0.18 \text{ h}^{-1}$, $\mu_{max} = 0.2 \text{ h}^{-1}$ and $K_s = 0.1 \text{ kg m}^{-3}$, the concentration of the substrate at the continuous fermenter outlet is 0.9 kg m^{-3}, which also happen to be the substrate concentration in the fermenter since its contents are well mixed. The outlet cell concentration, $x = (S_i–S)0.6 = (50–0.9)0.6 = 29.46 \text{ kg m}^{-3}$. The productivity of the fermenter is (see Eq. 5.38) $= Dx = 0.18(29.46) = 5.30 \text{ kg m}^{-3} \text{ h}^{-1}$. If $V \text{ m}^3$ is the fermenter volume, the production rate of yeast is DxV. The required production rate of the biomass is 25 tonnes per day or 1041.67 kg h^{-1}. Thus $DxV = 1041.67$ or $V = 1041.67/5.30 = 196.4 \text{ m}^3$. The amount of alkane consumed is $1041.67/0.6 = 1736.11 \text{ kg h}^{-1}$.

16. Consider a small element of the tube shown in Fig. 1a, which has a small volume dV in which the spore number concentration is N which changes by dN due to thermal death. If F is the volumetric flow rate of the liquid food through the tube, a spore mass balance can be written on the basis that the reduction in spore numbers is equal to the spore numbers perished in this section. Thus: $FdN = R_{death}dV = -2.5NdV$. In other words, $F\frac{dN}{dV} = -2.5N$, which is a first order differential equation that can be solved with the boundary condition, $N=N_i$ at $V = 0$, to yield a relationship between spore number and volume as: $F \ln\left(\frac{N_i}{N}\right) = 2.5V$. The residence time in the tube is given by $\tau = \frac{V}{F}$. Therefore $\ln\left(\frac{N_i}{N}\right) = 2.5\tau$, which can be used to estimate the residence time required for the spore load to fall from $N_i = 10^{11}$ to $N = 10^{-3} \text{ (m}^{-3})$ to give $\tau = 13.82 \text{ min}$. In

general, it follows that the spore numbers are related to the residence time by the eqn: $N = N_i e^{-2.5\tau}$.

If a spore mass balance is performed on the well mixed stirred vessel, employing the same notations, we have:

$$FN_i - FN_o = R_{\text{death}} V = 2.5 N_o V \text{ or } N_o = \frac{N_i}{1 + 2.5\tau}$$

Substituting the values of N_i and N_o given, $\tau \cong 4 \times 10^{13}$ min. Comparing the residence times in the tubular steriliser with no axial mixing with a well stirred steriliser, it is clear that mixing is most undesirable when such high values of conversion (14 orders of magnitude!) have to be achieved. Having said this, it is impossible to achieve perfect plug flow in a tubular reactor, and some axial mixing will be observed in practice, so time will have to be added to the residence time estimated of 13.82 min.

17. (a) Production cost per bottle of flavour C_0 = variable cost + fixed costs = $(47.73 + 0.1\ P^{1.2}) + (1750 + 7325)/P$. The production cost will be minimum when $dC_0/dP = 0$; and $d^2C_0/dP^2 > 0$. By equating the derivative to zero, we get $P = 165$; and the second derivative is also positive. In other words, 165 bottles have to be produced per day in order to minimise the cost per unit. The daily profit at this production schedule will be Sales revenue – production cost = $173(165)-[165(47.73 + 0.1\ (165)^{1.2} + 1750 + 7325] = £$ 11,911

(b) The daily profit $P_R = 173P - [(47.73 + 0.1\ P^{1.2})P + 9075]$. For the daily profit to be maximised, $(dP_R/dP) = 0$ and $(d^2P_R^2/dP) < 0$. By equating the derivative to zero, we get $P = 198$; and the second derivative is also negative. The daily profit will therefore be obtained by putting $P = 198$ in the expression for $P_R = £$ 4439.2.

(c) At break even, $P_R = 0$. We therefore have a nonlinear equation: $125.27P-0.1P^{2.2}-9075 = 0$. This can easily be solved either by using a software package or by using excel spreadsheet to plot P_R against P to yield $P = 87.33$ when $P_R = 0$. In other words, the breakeven production schedule will involve making 88 bottles per day.

18. (a) Mass of fermentation broth per cycle = density × batch size = 1te/m^3 × 100 m^3 = 100te. The number of production cycles per year = 330d ÷ 5d = 66. The mass of broth produced per year = mass of broth per cycle × 66 = 6600 te. The mass of product = mass of broth involved per year × 15% = 6600te × 15% = 990 te. The mass of beet molasses used = mass of product ÷ 50% = 1980 te. Therefore, the cost of beet molasses per year = 1980 te × 65 £/te = £128,700. The cost of chemicals, energy etc. per year at 30 £/te of beet molasses = 30× 1980 te = £59,400. It is also desirable to assume that labour is paid for all 365 days of the year, so the labour cost = 300£/d × 365 d = £109,500 per annum. Thus, the total operating expenditure per annum = £ 297,600. Since the capital is written down in 8 years, the capital investment made in the first year is used over

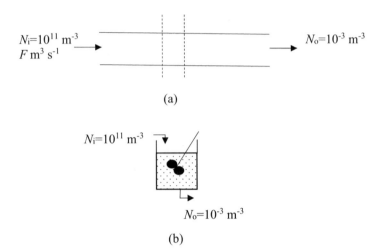

$N_i = 10^{11}$ m^{-3}
F m^3 s^{-1} $N_o = 10^{-3}$ m^{-3}

(a)

$N_i = 10^{11}$ m^{-3}

$N_o = 10^{-3}$ m^{-3}

(b)

Fig. 1 (a) Steady state plug flow tubular steriliser where the spore number concentration only changes along the length, and (b) steady state continuous well mixed vessel where the spore concentration is uniform throughout the vessel, which is also the same as the outlet concentration

7 years. Likewise, the capital investments made in the second and third years are used over 6 and 5 years, respectively. Therefore, the total capital cost per year may be considered to be equivalent to £100,000/7 + £150,000/6 + £200,000/5 = £ 79,286. Thus, the total production cost per annum = £(79,286 + 297,600) = £ 376,886, and the cost of producing the product is: 376886/990 = £ 380.69 per te. With a selling price of £400 per te, the process appears to be just about viable

(b) The cash flow and investment table over an 8 year duration of the project, without considering the time value of money and assuming that expenditure and revenues also do not change over this period (which is highly unrealistic) is as below:

Year	Investment capital (K)	Operating expenditure (C)	Revenue income (R)	Net balance (P)	Cumulative balance
0	0	0	0	0	0
1	100,000	0	0	−100,000	−100,000
2	150,000	0	0	−150,000	−250,000
3	200,000	0	0	−200,000	−450,000
4	0	297,600	396,000	98,400	−351,600
5	0	297,600	396,000	98,400	−253,200
6	0	297,600	396,000	98,400	−154,800
7	0	297,600	396,000	98,400	−56,400
8	0	297,600	396,000	98,400	42,000
9	0	297,600	396,000	98,400	140,400

(c) (i) The company is back in positive cash flow sometime between 7 and 8 years from the start. (this may seem too long from the point of view of an investment proposition!).

(c) (ii) The maximum negative exposure of the investment is at the end of year 3, which is to a tune of £450,000.

(d) (i) The net present value NPV can be estimated by noting that over the 15 years of project duration, actual production only takes place for 12 years since 3 years is taken up in constructing the plant. Therefore NPV = 400£/te × product output p.a. (te) × 12 - [100,000 × (1 + 8%)^7 + 150,000 × (1 + 8%)^6 + 200,000 × (1 + 8%)^5] - 297,600 × 12 = £477520.8

(d) (ii) The breakeven time is obtained by solving the equation we get by replacing 12 in the above expression by t and equating the expression to zero. This is a nonlinear equation and it can be solved to yield the breakeven time as 7.13 years. This is consistent with the cashflow and investment table, according to which the breakeven time is between 7 and 8 years.

(d) (iii) The breakeven price is obtained by solving the equation we get by replacing 400 by say, p, and equating the expression in Part d (i) to zero. This works out to £359.8.

(e) The internal rate of return is defined as the value of the discount rate at which the net present value becomes zero. This can be estimated by trial and error by assuming the discount to be X (i.e. replacing 8 in Part d). Doing so, results in the internal rate of return being 0.23 or 23%. Thus, the 8% discount rate is much lower and makes the investment viable.

Index

© Springer Nature Switzerland AG 2022
K. Niranjan, *Engineering Principles for Food Process and Product Realization*,
Food Engineering Series, https://doi.org/10.1007/978-3-031-07570-4

Printed in the United States
by Baker & Taylor Publisher Services